Advances in Modern
Environmental Toxicology

VOLUME II

OCCUPATIONAL HEALTH HAZARDS
OF SOLVENTS

Edited by

ANDERS ENGLUND
Bygghalsan, Construction Industry's Organization
for Working Environment, Safety and Health, Stockholm

KNUT RINGEN
Workers' Institute for Safety and Health, Washington, D.C.

MYRON A. MEHLMAN
Environmental and Health Sciences Laboratory
Mobil Oil Corporation, Princeton

Published by:
Princeton Scientific Publishing Co., Inc.
Princeton, New Jersey

Printed and bound in the United States of America.

PRINCETON SCIENTIFIC PUBLISHING CO., INC.
P.O. Box 2155, Princeton, New Jersey 08543

LIBRARY OF CONGRESS 82-62164   ISBN 0-911131-02-7

Advances in Modern Environmental Toxicology Series: ISSN 0276-5063
                                                      ISBN 0-911131-00-0

Cover illustration: From N. Krivanek, p. 11.

TABLE OF CONTENTS

PREFACE v
ACKNOWLEDGMENTS vii

SECTION I

CHAPTER 1   The Toxicity of Paint Pigments, Resins, Drying Oils
and Additives
Neil Krivanek 1
CHAPTER 2   Solvent Technology in Product Development
Charles Hansen 43
CHAPTER 3   A Comparative Toxicological Evaluation of
Paint Solvents
Myron Mehlman and Charles Smart 53
CHAPTER 4   The Influence of Environmental Factors on Research
and Development of Paint Products
Lennart Dufva 69

SECTION II

CHAPTER 5   Experimental Approach to the Assessment of the Car-
cinogenic Risk of Industrial Inorganic Pigments
Cesare Maltoni, Leonildo Morisi and
Pasquale Chieco 77
CHAPTER 6   Chemical Hazards in Painting in the Construction
Industry
Riitta Riala 93
CHAPTER 7   Early Results of the Experimental Assessment of the   77
Carcinogenic Effects of One Epoxy Solvent:
Styrene Oxide
Cesare Maltoni 97
CHAPTER 8   Health Hazards Among Painters
Knut Ringen 111

SECTION III

CHAPTER 9   Work Load and Uptake of Solvents in Tissues of Man
             Irma Åstrand 141
CHAPTER 10   Respiratory Symptoms Due to Paint Exposure
             Kaye Kilburn 153
CHAPTER 11   Organic Solvents and Kidney Function
             Alf Askergren 157
CHAPTER 12   Cancer Incidence and Mortality Among Swedish Painters
             Anders Englund and Goran Engholm 173

             INDEX 187

SECTION IV: REPRINTS

Long-term Exposure to Jet Fuel
Bengt Knave et al. 197
Exposure to Organic Solvents
Stig-Arne Elofsson et al. 225

PREFACE

In 1975, the International Federation of Building and Wood Workers (IFBWW) organized a conference of trade unionists and interested experts to explore occupational health hazards from solvents and other paint components. The outcome was published as a small but important volume entitled Occupational Health Hazards in the Painting Trades (IFBWW, Geneva Switzerland, 1975). Two significant conclusions can be drawn from this conference: first, that there is widespread concern about hazards to workers' health caused by the use of solvent-based materials; second, that the available information on the subject was very limited.

The papers contained in this volume attempt to address these matters. They were first presented in Stockholm at the "International Symposium on Occupational Health Hazards Encountered in Surface Coating and Handling of Paints in the Construction Industry." In late 1981 and early 1982, they were revised and updated for inclusion in this book. Much of this revision stems from the exchanges that were stimulated by the symposium, both during and subsequent to the meeting at Stockholm.

The contributors to this volume are primarily Western European and North American representatives from labor, industry, government and academia. Their presentations include the latest available data on the toxicity of the various substances used in solvent-based materials, as well as expert scientific discussions of the health hazards presented by these substances.

Since the IFBWW conference, major advances have been made in the level of our knowledge of these health hazards. This volume represents the first thorough compilation of advances.

Myron A. Mehlman

ACKNOWLEDGMENTS

The Symposium was organized by Bygghalsan—The Construction Industry's Organization for Working Environment, Safety and Health. We are particularly grateful to Kurt Mansson, Director of Bygghalsan, and Margaret Hogberg and Lolo Heijkenskjold for their careful attention to every detail.

Consulting in the planning and organization of the Symposium were the Workers' Institute for Safety and Health, Washington, D.C., the Environmental Sciences Laboratory, Mount Sinai School of Medicine, New York, International Section on the Prevention of Occupational Risks in the Construction Industry—I.S.S.A., the Swedish Plastics and Chemical Suppliers Association, Swedish Paint and Printing Ink Manufacturers' Association. Financial support for the symposium was provided by the Swedish Work Environment Fund and the Swedish Council for Building Research.

Partial support for this volume from International Brotherhood of Painters and Allied Trades of Washington D.C. and Environmental Protection Agency, Research Triangle Park, N.C. is acknowledged.

SECTION I

CHAPTER 1

THE TOXICITY OF PAINT PIGMENTS, RESINS, DRYING OILS, AND ADDITIVES

Neil Krivanek, Ph.D.
Haskell Laboratory
Central Research and Development Department
E. I. du Pont de Nemours & Company
Wilmington, Delaware 19898

INTRODUCTION

Paint protects surfaces from adverse exposure conditions and also helps give the coated object an aesthetically attractive appearance. In order for a surface coating to function as an effective barrier against adverse conditions, it should not react with the external agents. Building inertness into a film is a basic goal of paint technology. The purpose of this paper is to review the toxicity of paint pigments, binders, which includes resins and drying oils, and additives. Since the paint industry uses several thousand different chemicals in producing its products, all materials cannot be covered. This paper focuses on the commonly used chemicals of these three categories.

The large number of ingredients and the large number of different formulations, including colors, makes it clear that discussion of paint toxicity is best accomplished by discussing toxicity of individual ingredients rather than individual formulations. Before discussing specific ingredients, the major chemical and biological properties of the finished products, the dried or cured paint coating, are listed.

- The coating is low in chemical reactivity.
- Most of the non-volatilized paint components are bound within the coating to give the film its desired properties.
- The cured coating is a high molecular weight polymer. In general these polymers are poorly absorbed by the gastrointestinal tract and skin.[1,2,3,4] Inhalation of fully cured coating has not been associated with unusual health hazards, but there is little experimental data to support this claim.

Toxicity problems that are encountered with cured coatings are most often associated with chemicals that can leach out of the coating in body fluids. In certain circumstances, such as fires, thermal decomposition products of the coating can be inhaled and produce toxic effects.

The toxicity of liquid or uncured paint is associated with individual paint components and contaminants. It is mainly during the application of paint that the potential for overexposure to these ingredients is greatest. Potential human health hazard from paint is a complicated problem which involves the amount of substance in paint, type of application (spray, brush or roller) general ventilation, protective equipment and personal work habits. The remaining discussion of this paper is on toxicity.

Historically, paints have always been associated with health risk. This was mainly due to the presence of the heavy metal pigments of lead, cadmium, and chromium. The introduction of synthetic chemicals, both inorganic and organic, in the 1920's increased the number of chemicals the painter encountered; and the introduction of the spray application technique increased the potential for human exposure, especially by the inhalation route. Changes in paint formulation occur constantly and in many cases a less toxic ingredient has replaced a more toxic one. Two examples of this kind of substitution are the use of titanium dioxide instead of white lead pigments and replacement of many organic solvent-based consumer paint formulations with water-based formulations.

PIGMENTS

Generally, pigments are finely divided solids which give paint color, hiding power, durability, and film hardness. Pigments constitute 20–60% by weight of a paint. An ideal property of a pigment is that it should not be soluble in the paint and should not react with other paint constituents, unless specifically engineered for that purpose. Inorganic pigments have greater usage than the organic pigments and are discussed first.

The physical properties of pigments such as particle shape and size vary. The most common shape is spherical, such as titanium dioxide; other pigments are needle shaped, e.g., zinc oxide; or plate-like, e.g., mica and talc.

Particle size generally has an upper limit of about 44 microns mean diameter and a lower limit of 0.01 micron, especially seen with the carbon blacks. Particles less than 44 microns in diameter will pass through a number 325 mesh screen. Particles larger than 30–40 microns mean diameter often make the coating too coarse and coarseness is considered undesirable in most cases. Generally the size of pigment particles is less than 3 micron in diameter. Titanium dioxide particles have an average size of 0.2 micron in diameter.[5]

The particle size of dry pigment powders are in the range of respirable dust, 10 to 0.5 micron; thus, inhalation of dry pigments can present a significant inhalation health hazard to the persons handling these materials.

Once dry pigments have been incorporated into paint, their potential health hazard decreases. The method of paint application can significantly modify the health hazard potential of paint. Brush or roller application produces small amounts of aerosolized, respirable paint so that the inhalation health hazard for the paint pigment remains low. Spray application of paint produces significant amounts of aerosolized, respirable paint particles, and not only increases hazard potential for pigments, but also for solvents, resins, and additives. There is a paucity of published experimental data on pulmonary responses to aerosolized paint particulates.

The toxicity of commonly-used paint pigments is summarized in Table 1.

White pigments constitute over 90% of all pigments used. In the past, these pigments were lead compounds. Many lead compounds are very toxic and many persons have suffered lead poisoning mainly from inhalation of dry pigment dust during paint manufacture. Inhalation of lead pigments can also occur during removal of lead-based coatings, especially by dust generating mechanical techniques such as sanding. In homes painted with lead pigment formulations, children have suffered lead poisoning because they ate paint chips or gnawed on objects coated with paints containing lead pigments.

Lead pigments are no longer used in home construction paints. Titanium dioxide has replaced these white lead pigments. Titanium dioxide has very low oral, dermal, and inhalation toxicity. Chronic oral studies in animals show no increase in cancer due to ingestion of titanium dioxide dust.

Other white pigments, such as calcium carbonate, barium sulfate, and aluminum silicate are low in toxicity.

The silicas and silicates comprise a group of mineral dusts which have no significant oral and dermal toxicity. However, inhalation of crystalline silica dust (quartz, cristobalite, or tridymite) can produce silicosis, a disabling lung condition characterized by fibrosis of pulmonary tissue which can eventually be fatal. Safe exposure limits for these dusts are based upon their quartz content, the greater the amount of quartz, the lower the exposure limit.[6]

Diatomaceous earth is a natural material composed mainly of amorphous silica with variable trace quantities of crystalline silica. Diatomaceous earth is less potent than crystalline silica in producing a damaging pulmonary response, and its exposure limit reflects this difference. Amorphous silicas are considered to be much less fibrogenic to the lung than the crystalline types.

The talcs are divided into two forms—fibrous and non-fibrous. The fibrous form can produce a response in the lung which is similar to that seen for asbestos. Asbestos-type particles have been identified as impurities in talcs and the American Conference of Governmental Industrial Hygienists exposure limit, i.e., Threshold Limit Value (TLV®)[7] for fibrous talc is the same as for asbestos. Non-fibrous talc is much lower in pulmonary toxicity than the fibrous form. The available human and animal data on the carcinogenic potential of talcs are controversial and difficult to evaluate because of the presence of fibrous minerals; such as asbestos, and the difficulty of accurately identifying fibers in talcs. Because of the latter uncertainty, the non-fibrous forms of talc have been used to replace talcs which contain mineral fibers.

Repeated and prolonged exposure to high concentrations of mica may cause a disabling pneumoconiosis. Exposure to air levels of mica below 20 million particles/ft.[3] of air is considered sufficient to prevent pneumoconiosis development.[6]

TABLE 1

TOXICITY SUMMARY OF PIGMENTS COMMONLY USED IN PAINTS FOR THE CONSTRUCTION TRADE

Chemical Name Structure Physical State	Occupational Exposure Limit	Toxicity
INORGANIC PIGMENTS		
Aluminum powder	OSHA* - N.A.**	Aluminum metal powder has no appreciable toxicity by the oral and dermal route. Intratracheal injection of large doses (100 mg) in rats produced focal pulmonary fibrosis, but this response did not occur in identically treated hamsters. Hamsters and guinea pigs exposed up to 100 mg/m^3 aluminum powder, 6 hrs/day for 6 months developed transient alveolar proteinosis; no fibrosis occurred.[19] No clear-cut evidence of pulmonary damage has been found in occupational exposure to aluminum dust alone, but cases of pulmonary fibrosis have been reported in workers exposed to aluminum flake powders coated with lubricants, mineral oils or stearin.[20,21,22]
Al metal	TLV***–10 mg/m^3 Aluminum pyropowders	
Silvery powder	TLV–5 mg/m^3	
Aluminum silicate	OSHA–N.A.	Oral ALD (rabbit) is greater than 11,000 mg/kg. Not a skin irritant. Very mild transient eye irritant. LC50 (rat, 4 hr.) is 2.3 mg/L. Intratracheal insufflation of 1 mL of a 5% suspension in saline produced a fibrotic response in rat lung.[23] A single intrapleural injection of 20 mg into rats produced 10% mesotheliomas, suggesting solid-state carcinogenesis.[24]
Al$_2$O$_3$–SiO$_2$2H$_2$O	TLV–10 mg/m^3 total dust <1%	
China clay, kaolin, white powder	quartz$_3$ or, 5 mg/m^3 respirable dust.	

*OSHA—U.S. Occupational Health and Safety Administration, 8 hour time-weighted average.
**NA.—Not assigned.
***TLV—Threshold Limit Value, American Conf. of Governmental Ind. Hygienists, 8-hour time weight average.

(Table 1 continued)

Antimony oxide Sb_2O_3 White powder	OSHA– 0.5 mg/m³ (as Sb) TLV– 0.5 mg/m³ (as Sb)		Oral LD50 (rat) is greater than 20 g/kg, low order of skin toxicity, not a skin sensitizer. Irritating to rabbit eyes, but not to humans. Inhalation of Sb_2O_3 by animals at a concentration of 2.76 mg/L produced no toxic effects.[25,26,27]
Fire-retardant pigment			
Barium sulfate $BaSO_4$ White insoluble powder	OSHA– 0.5 mg/m³ (as Ba) TLV–0.5 mg/m³ (as Ba)		No apparent toxicity by oral route, ALD is greater than 5000 mg/kg.[28] Used internally in humans as radiopaque material. Cleared as food packaging pigment. Accidental ocular contact caused little or no toxic effect.[29] Intratracheal injections in rats caused no effect.[30] Inhalation exposure of rats to 40 mg/m³ for 2 months showed initial pulmonary reaction—alveolar macrophage proliferation, which regressed during remainder of exposure. After exposure no lung damage was observed.[31] No evidence of carcinogenic potential in feeding or inhalation study in mice and monkeys.[31,32] Subcutaneous implants of polyethylene containing $BaSO_4$ showed solid-state carcinogenesis.[33] No effect on reproduction was seen in three generation mouse and monkey feeding study.[32]
Calcium carbonate $CaCO_3$ White powder	OSHA–N.A. TLV–10 mg/m³		Acute toxicity is very low. Used therapeutically in animals, as antacid.[34] Considered as nuisance particulate when airborne.[35]
Carbon black Black particles	OSHA– 3.5 mg/m³ TLV– 3.5 mg/m³		Oral ALD (rat) is greater than 25,000 mg/kg.[36] No significant pathological finding upon subcutaneous or intraperitoneal injection in mice.[37] Some carbon blacks such as lamp black or furnace black made from oil may contain small amounts of

(Table 1 continued)

50–300 Å in diameter	polycyclic hydrocarbons, such as 3,4-benz(α)pyrene, as known carcinogen. Thus, in long-term feeding and skin painting studies in mice with carbon black containing polycyclic hydrocarbons, or extracts of these carbon blacks, a few tumors were produced.[9] Chronic inhalation studies in monkeys, hamsters, mice, guinea pigs, and rabbits at concentrations between 50 and 100 mg/m^3 did not produce any excess cancers.[38] Epidemiological studies with carbon black workers have not shown any increased cancer risk.[10]
Copper (I) oxide Cu$_2$O Red powder	OSHA– 1.0 mg/m^3 (as Cu) TLV–1.0 mg/m^3 (as Cu)	Oral LD50 (rat) is 470 mg/kg.[39] Toxicity due to copper content. No evidence of carcinogenic potential via implantation study.[40]
Copper (II) oxide CuO Black powder		
Anti-fouling pigment		
Iron oxide Fe$_2$O$_3$ (red) haematite Fe$_3$O$_4$ (black)	OSHA–10 mg/m^3 (as a fume) TLV–5 mg/m^3 (as iron oxide fume)	Low in oral toxicity, lethal dose for rats about 1000 mg/kg. Intraperitoneal injection of iron oxide or subcutaneous implantation of ferric oxide discs did not produce tumors.[41,42] Hamsters given intratracheal injection of Fe$_2$O$_3$ up to 50 mg per dose showed no significant increase in respiratory tract tumors.[43]

magnetite

Fe_2O_3 (brown)
Colored powder

Hamsters exposed to 40 mg/m³ Fe_2O_3 dust daily for their lifetime showed no observable respiratory tract tumors.[44] Humans exposed (14 to 16 years) to fumes (1.3–294 mg/m³) of iron oxide with some chromium oxide and nickel oxide showed radiological evidence of mixed dust pneumoconiosis without fibrosis.[45] Another study of workers exposed to pure Fe_2O_3 dust showed nodular opacities in chest radiographs, but no fibrosis.[46] These results support the concept that pure iron oxide does not produce fibrotic pulmonary effects, but inhalation of iron oxide dust with other substances may cause damage.

Lead chromate
$PbCrO_4$

Yellow powder

OSHA–
0.05 mg/m³
(as Pb)
OSHA–
0.1 mg/m³
(as CrO_3)
(Ceiling)

TLV–
0.05 mg/m³
(as CR)

Intraperitoneal LD75 (guinea pig) is 156 mg/kg.[47] Intratracheal implant of calcium chromate produced pulmonary tumors in rats.[48] Subcutaneous injection of lead chromate in rats produced local sarcomas.[49] Epidemiological studies on workers in the chromate producing industry showed an increase in lung cancer.[50,51]

The exact chemical form of the chromate carcinogen has not been established.

Lead tetraoxide
2 Pb0.PbO₂
(Pb_3O_4)
Orange-red powder
lead

OSHA–
0.05 mg/m³
(as Pb)
TLV–
0.15 mg/m³
(as Pb)

Intraperitoneal LD50 (guinea pig) is 220 mg/kg.[52] Over-exposure to lead can produce a variety of systemic toxic effects: depression of the central nervous system, anemia, gastrointestinal upset and degeneration of liver and kidney tissue. Inhalation is the most important route of occupational exposure for inorganic lead compounds. Toxic effects are generally not manifested immediately, but require time for body

(Table 1 continued)

Mica Potassium aluminum silicate $K_2O \cdot 3Al_2O_3$ $6SiO_2 \cdot 2H_2O$ Plate-like particles	OSHA–20 mppcf* TLV-20 mppcf* (<1% quartz)	lead burden to build up to critical concentrations in the particular organ.[53] No acute toxicity recognized. Inhalation of respirable particulate above 20 mppcf* produced mild pneumoconiosis in humans.[54]
Silica SiO_2 Crystalline form Quartz, cristobalite, tridymite	Dependent upon crystalline SiO_2 content OSHA– $\dfrac{10 \text{ mg/m}^3}{\% \; SiO_2 + 2}$	No appreciable toxicity by oral or dermal route. Inhalation of the crystalline forms can cause fibrogenic pulmonary response, which can be disabling and is termed silicosis.[55]
White powder	TLV– $\dfrac{300 \text{ mppcf*}}{\% \text{ quartz} + 10}$ TLV– $\dfrac{10 \text{ mg/m}^3}{\% \text{ respirable quartz} + 2}$ TLV–$\dfrac{30 \text{ mg/m}^3}{\% \text{ quartz} + 3}$	

*million particles per cubic foot

Amorphous form SiO_2	OSHA–20 mppcf* TLV–5 mg/m³, total dust TLV–2 mg/m³, respirable dust (<5 uM)	Inhalation of non-crystalline silica does not produce a fibrogenic pulmonary response in guinea pigs at 60 mg/m³.[56] Occupational exposure to amorphous silica has produced some possible cases of silicosis.[55,57]
Diatomaceous earth, natural amorphous SiO_2	TLV–1.5 mg/m³, respirable dust	
Colorless to gray powder		
Talc Magnesium silicate (hydrated) $3\ MgO \bullet 4\ SiO_2 \bullet H_2O$ White powder	OSHA–20 mppcf* (non-asbestiform) TLV–20 mppcf* (non-asbestiform)	No acute toxicity known. Intratracheal and intrapleural injection of talc into laboratory animals has produced fibrotic and granulomatous responses. Inhalation exposure of animals and humans to talc can produce pneumoconiosis after chronic exposure, this can progress to fibrosis and granulomatosis. The existence of fibrous components and asbestos-like materials in talc has tied talc exposures to materials with carcinogenic activity. The nature of the carcinogenic potential of talc is unclear due to the existence of both fibrous and nonfibrous forms and presence of impurities such as asbestos, a known animal and human carcinogen.[57]
Talc (fibrous)	TLV–2 fiber/cc (fibrous)	
Titanium dioxide TiO_2	OSHA–15 mg/m³ TLV–10 mg/m³	Very low oral toxicity, oral ALD (rat) is greater than 24,000 mg/kg.[58] Not a skin irritant or sensitizer.[59] Intratracheal injection in rats produced a variable response: no to mild macrophage infiltration in aveoli and transient alveolar wall thickening. Reaction typical of foreign body response to inert material.[60] Chronic inhalation exposure in rats for 13 months up to 328 mppcf*
White powder		

*million particles per cubic foot

(Table 1 continued)

		showed only slight lymphoid proliferation.[60] A long-term ingestion study has been done and no evidence of excessive tumors has been found.[61] Human experience to only TiO_2 exposure has not shown significant pulmonary alterations.[62]
Zinc oxide	OSHA–5 mg/m^3	Negligible oral toxicity. Used in topical skin creams, negligible dermal toxicity, not a skin irritant. Occupational exposure to zinc oxide fumes in humans can produce a transitory illness characterized by fever, chills, nausea and general discomfort.[63]
ZnO	TLV–5 mg/m^3	
White powder		
Zinc dust	OSHA–N.A. TLV–N.A.	No known acute or chronic toxicity in animals or man. Zinc is an essential nutrient for man. Heating of zinc dust can produce zinc oxide fumes which has known toxicity (see Zinc oxide).
Zinc chromate $ZnCrO_4$ also contains K_2CrO_4, ZnO $Zn(OH)_2$	OSHA– 0.10 mg/m^3 (as CrO_3) TLV–0.05 mg/^{3} (as Cr)	Oral ALD (rat) is 200 mg/kg. Skin irritant and sensitizer in humans. Intramuscular implant of basic zinc chromate has produced local sarcomas in rats. Chromium (VI) compound are carcinogenic in animal studies. Occupational exposure data shows excess lung cancer in chromate production workers.[64,12] Spray painters showed an excess of respiratory cancer, but other exposures could not be discounted.[65]
Yellow powder		

ORGANIC PIGMENTS
Phthalocyanine blue

OSHA–N.A.
TLV–N.A.

C.I. 74160*

Oral LD50 (rat) is greater than 17 g/kg. It was not a skin irritant when tested on guinea pig skin or mutagenic in the bacterial mutagen assay.[66] In an 8-month chronic study, subcutaneous injection in mice, 0.5 mg, produced no significant toxic effects.[67] Phthalocyanine green (C.I. 74260) is also negative in bacterial mutagen assay.[66]

(Table 1 continued)

Toluidine Red, C.I. Pigment Red 3, C.I. 72110	OSHA–N.A. TLV–N.A.	Oral ALD (rat) is greater than 7500 mg/kg. Rats given 2500 mg/kg/day for two weeks (10 doses) showed no toxic effects upon gross and microscopic examination.[68]
Red powder C.I. Pigment Red 88	OSHA–N.A. TLV–N.A.	Oral LD50 (rat) is greater than 16 g/kg. Not known to be a skin irritant.[69]
C.I. 73312		
Red powder C.I. Pigment Violet 19	OSHA–N.A. TLV–N.A.	Oral ALD (rat) is greater than 7500 mg/kg. A 10% paste was slightly irritating to guinea pig skin and it was not a skin sensitizer for guinea pigs.[70]
C.I. 46500		

Violet red powder

C.I. Pigment Red 49

Lithol red, sodium salt

C.I. 15630

Red Powder

OSHA–N.A.
TLV–N.A.

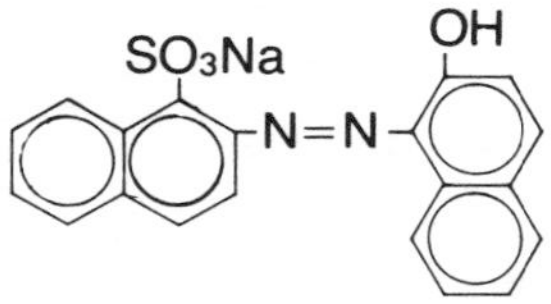

Negative in bacterial mutagen assay.[14]

para-Nitroaniline

OSHA–1 ppm
TLV–1 ppm

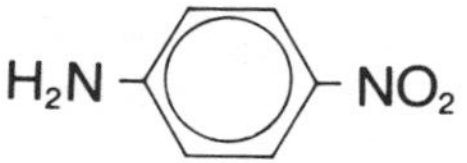

C.I. 37035

Yellow powder

Oral LD50 (rat) is 3249 mg/kg.[71] Readily skin absorbed. In 10-day oral route study in rats at 675 mg/kg mortality and kidney damage occurred.[72] Inhalation of 5 mg/m^3 at 5 hours/day for 4 months decreased hemoglobin levels in test animals.[73] Occupational exposures have produced toxic effects—weakness, headache and methemoglobinemia.[74]

The two most common metallic dusts used in paint are aluminum powder and zinc dust. Aluminum powder and zinc dust are low in toxicity by normal routes of exposure: ingestion, inhalation and skin contact.

Zinc oxide is a white pigment and is low in toxicity. However, inhalation of freshly-formed zinc oxide fumes, such as those produced by heating zinc metal to high temperatures, can produce a transient set of symptoms known as metal fume fever characterized by chills, fever, nausea, muscular pain, and headache.

Black colored pigments are second only to the whites in terms of paint pigment usage. Carbon black is the most common black pigment and is produced from incomplete combustion of petroleum gas. Lamp black, the other major black pigment is made from incomplete combustion of oil. Certain carbon blacks are cleared as indirect food additives by the Food and Drug Administration. The toxicity of carbon black is negligible by the oral route. Inhalation of extremely high concentrations of carbon black has produced pulmonary damage in experimental animals.[8] Carbon blacks which contain known amounts of polynuclear aromatic hydrocarbons, such as 3,4 benz (α) pyrene, have produced tumors in animals.[9] Carbon blacks which contain non-detectable amounts of polynuclear aromatics have not produced excess tumors in laboratory animals when tested by various routes of administration.[9] Epidemiological studies of carbon black workers shows no excess cancers.[10]

Iron oxides are found in red, brown and black inorganic pigments. Their oral and dermal toxicity is low. Inhalation of iron oxides can produce changes on pulmonary x-rays which resemble those seen in silicosis but there is no fibrosis and the condition, termed siderosis, is not debilitating.

The most popular yellow inorganic pigments are lead and zinc chromates, although lead chromate is no longer used in consumer paints. Zinc chromate and lead chromate are moderately toxic by the oral route. Chromates (as Cr VI) are skin sensitizers.[11] The major concern of overexposure to chromates stems from epidemiologic studies which report excess lung cancer of workers in the lead and zinc chromate producing industry. Results from the chromate user industry are not as clear-cut as those in the chromate production. Chromates have produced tumors in laboratory animals at the site of administration: intratracheal, subcutaneous or intramuscular injection.[12]

Lead tetraoxide is a red colored oxide, also known as red lead, (Pb_3O_4) and is very toxic.[53] Overexposure can produce systemic lead poisoning. This pigment is used in large quantities on iron and steel because of its corrosion inhibiting properties.

Organic pigments. There are hundreds of organic pigments available for use by the paint industry. In view of this, a representative selection of some commonly used organic pigments is discussed. Acute toxicity is low for most organic pigments. Long-term effects in man have not been well studied. Some aromatic azo dyes have been tested in chronic animal studies.[13]

Reports have been published on the mutagenicity of organic colorants. In one study, 19 colorants were tested for mutagenicity in bacterial mutagenic assays.[14] Two dyes; Para red, C.I. 12070, and Dinitroaniline orange, C.I. 12075, were weakly positive. The other 17 were negative. They were Alkali blue, C.I. 42750; Aluminum, elemental; Cadmium red, C.I. 77196; Diarylide orange; C.I. 21090; Fire red, C.I. 12085; Hansa yellow, C.I. 11680; Iron blue, C.I. 77510; Lithol red, C.I. 15630; Lithol rubine, C.I. 15850, Molybdate orange, C.I. 77605, Naphthol red, C.I. 12315; Phthalocyanine blue, C.I. 74160; Phthalocyanine green, C.I. 74260; Red 2B, C.I. 15865; Red lake C red, C.I. 15585; and Rhodamine, C.I. 45160. The significance of positive bacterial mutagenicity results in terms of human risk is not fully known.

The acute toxicity of the following organic pigments is low: Pigment violet 19, C.I. 46500; Pigment red 3, C.I. 12110; Pigment red 49, C.I. 15630; Pigment red 88, C.I. 73312; and Phthalocyanine blue, C.I. 74160. Para-nitro aniline, C.I. 37035, is capable of producing toxic effects on blood, mainly methemoglobinemia.

RESINS AND DRYING OILS

The vehicle or nonpigment portion of paints contains components collectively termed binders. Their function is to provide the chemistry for holding the paint film itself together and holding the paint to the substrate, i.e., item to be painted. Almost all binders in modern paints, varnish and lacquer films are composed of polymeric materials—resins and drying oils. Resins are amorphous and describe a physical state rather than an exact chemical entity. The main functions of resins are to provide film hardness, gloss, surface adhesion, and resistance of the film to adverse conditions, such as alkali, acids, and other harsh environments. The drying oils form films, i.e., polymers, by reacting with oxygen from the air after they are applied to the substrate.

Resins are derived from either natural sources or are chemically synthesized. Natural resins have been used in paint for centuries while the synthetics have been commercially available since the early 1900's.

Resins are low in volatility, do not have a fixed melting point, and are generally soluble in organic solvents, but insoluble in water.

Resins are polymers or polymer precursors. In liquid paint the molecular weights of resins are in the low thousands, but in the cured coating molecular weights in the millions are not uncommon. Polymers are composed of repeating units of smaller molecules (monomer) linked together by chemical bonds. Polymers are formed by two major reaction processes—addition or condensation polymerization. In an addition reaction, a free radical is formed by decomposition of an initiator, such as benzoyl peroxide, which opens one of the monomer's double bonds or a ring and forms the polymer. No reaction products are eliminated. Addition reactions are involved in polymerization of vinyl chloride, ethylene, propylene, styrene, methyl methacrylate, tetrafluoroethylene, and other olefins.

The other type of polymerization is condensation polymerization. In this case the polymer is formed by elimination of a reaction product, often water. Condensation polymerization reactions are found with polyamides, polyesters, alkyds, and silicones.

Once formed from monomer the polymer is distinctly different from the monomer in physical, chemical, and toxicological properties.

There are many different types of resins used by painters in the construction industry and they include alkyd, acrylic, vinyl, epoxy, urethane, phenolic, cellulosic, chlorinated rubber, styrene-butadiene, polyester and silicone types. In many cases mixtures of different synthetic resins are incorporated into a paint to provide certain properties not obtained in a paint based on only one resin. The amount of resin in paint varies, but values in the 20–35 weight percent range are common. The published literature contains little toxicity information on resins. The toxicity of common resins and drying oils is summarized in Table 2.

Natural Resins. An important natural resin is made from rosin. Rosin is a pine tree extract and has been used in paints for many years. Two main types of rosin exist—gum rosin and wood rosin. Gum rosin is obtained from live tree sap. Gum rosin is composed of about 70% rosin, 20% turpentine and 10% water. Wood rosin is obtained from tree stumps via solvent extraction processes.

Rosin is about 85% rosin acids and 15% neutral substances. These rosin acids all contain a phenanthrene backbone with various side groups. Abietic acid is the most common. Rosins are used mainly in alkyd paints.

The toxicity of natural resins is low. Abietic acid is low in toxicity, but undiluted it may be irritating to eyes, skin, and upper respiratory tract.

Rosins by themselves do not make a high quality paint, therefore, they are upgraded via chemical reactions. Some examples of these reaction products are: limed rosin (calcium rosinate), estergum, esterication of rosin with glycerol, and a variety of other products which can be formed when rosin is reacted with pentaerythritol, trimethylolpropane, phthalic anhydride, maleic anhydride, adipic acid or sebacic acid. The toxicity of these reaction products is low.

Synthetic Resins. The synthetic resins are made from monomers. Monomers are generally considered chemically reactive substances. Conversion of monomers into resins produces a material which is much less reactive than the monomer, but still retains some of the active groups of the monomer. Generally there is only a small amount of monomer starting material in the resin. Present technology uses manufacturing specifications where often at least 99% of the monomer is converted into oligomers and polymers. In cases where the monomer content may represent a health hazard with use of the polymer, the residual monomer content is reduced to only a few parts per million in the resin, so that in the final paint formulation there is generally no significant exposure to monomer. Two examples of this are residual vinyl chloride in vinyl polymers and epichlorohydrin in epoxy resins.

TABLE 2

TOXICITY SUMMARY OF RESINS COMMONLY USED IN PAINTS FOR THE CONSTRUCTION TRADE

Chemical Name Structure Physical State	Occupational Exposure Limit	Toxicity
Alkyd A polyester produced from reaction of polyhydric alcohol and poly-functional acid (glycerol and phthalic anhydride). Molecular weight varies with product.	OSHA–N.A. TLV–N.A.	Toxicity is low. Polymers are generally of high molecular weight and not readily incorporated into organism by oral or dermal route.
Alkyd resin surface coating based on dehydrated castor oil cross-linked with melamine-hexameth-ylol methyl ether.		Inhalation LC50 (rat, 4 hr.) is greater than 3625 mg/m$^{3.75}$
Amino-Urea (dimethylolurea and formaldehyde)	OSHA–N.A. TLV–N.A.	Uncured resin seldom encountered in construction industry as coating.

$$\left[\begin{array}{c} CH_2\,HN \quad\quad HN-CH_2 \\ \quad C=O \quad\quad C=O \\ \quad O-CH_2-N-CH_2-OH \end{array} \right]_n$$

(Table 2 continued)

Molecular weight varies

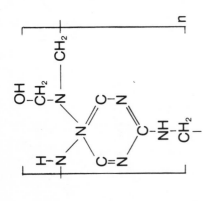

Polyacrylate coating, water reducible, crossed-linked with melamine formaldehyde resin.

Inhalation LC50 (rat, 4 hr.) is greater than 1694 mg/m^3.[75]

Chlorinated rubber

OSHA—N.A.
TLV—N.A.

Oral LD50 (guinea pig) is greater than 10 g/kg. Not a skin irritant or sensitizer in humans. Slight eye irritant. Carbon tetrachloride is used as a processing solvent and is retained in the rubber, about 8%.[76]

Cl content variable (65—67%)
Parlon®
White powder

Epoxy	OSHA–N.A. TLV–N.A.		Uncured epoxy resins can produce contact sensitization, cured resins do not. Liquid resins tend to be more toxic than solid resins.[17]
Epon® 1001, mw 1800, n=3 Solid	OSHA–N.A. TLV–N.A.		Oral LD50 (rat) is greater than 30 g/kg, moderate eye irritant. Not a skin irritant.[17]
Epon® 828, mw 370, n=1 Liquid	OSHA–N.A. TLV–N.A.		Oral LD50 (rat) is 11.4 g/kg, dermal LD50 (rabbit) is greater than 20 mL/kg, slight skin irritant, strong eye irritant. Repeated skin contact caused no skin tumors in animals. Variable results in bacterial mutagen assays.[17]
Two component epoxy coating polyepoxide resin cross-linked with polyamide.			Inhalation LC50 (rat, 4 hr.) is greater than 3375 mg/kg.[75]

(Table 2 continued)

Butyl glycidyl ether (reactive epoxy diluent) $C_4H_9OCH_2CH\text{-}CH_2$ (epoxide)		OSHA–50 ppm TLV–50 ppm	Oral LD50 (rat) is 2.26 g/kg. Mild eye and skin irritant. Skin sensitizer. Positive in mutagenicity tests. Testicular effects at 300 ppm in animals.[77]
Epichlorohydrin $CH_2\text{-}CH\text{-}CH_2Cl$ (epoxide) Colorless liquid Occasionally present in trace quantities in epoxy resins.		OSHA–5 ppm TLV–2 ppm	Skin sensitizer. Subacute inhalation exposures of rats at 9 ppm did not produce toxic effects.[78] Chronic inhalation exposure of rats produced tumors at 100 ppm and 30 ppm.[18]
Nitrocellulose $[\ CH_2\,ONO_2,\ ONO_2,\ H,\ H,\ ONO_2\ ]_n$ Variable degree of nitration Molecular weight range 10,000–150,000		OSHA–N.A. TLV–N.A.	Oral LD50 (rat) is greater than 5000 mg/kg.[79] In a limited chronic feeding study in mice receiving 0.2% nitrocellulose in diet as a nitrocellulose coated cellophane, no harmful effects were observed.[80]

Phenolics

OSHA–N.A.
TLV–N.A.

Condensation polymers of modified phenols with aldehydes

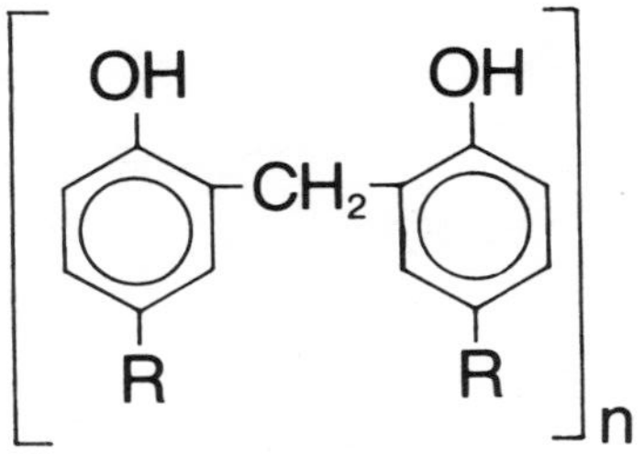

R = alkyl group,
t-butyl, octyl,
t-amyl

Phenolic resin-Novolac®

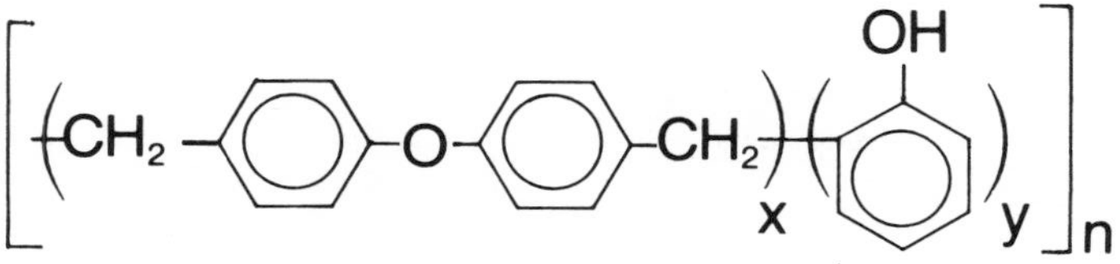

Oral LD50 (rats) is greater than 2000 mg/kg. Mild eye and skin irritant. No health effect from skin absorption or single exposure to vapors generated at room temperature or at 100°C.[81]

(Table 2 continued)

CMPDO®-25 and phenol Diphenyloxide modified Novolac® mw about 1000 Versamid® 100	OSHA–N.A. TLV–N.A.	Oral LD50 (rat) is 34.6 g/kg. Skin LD50 (rabbit) is greater than 6.8 g/kg. Slight skin irritant.[84]
Polyaliphatic amine and dimerized vegetable oil amine in epoxypolyamide resin.	OSHA–N.A. TLV–N.A.	The U.S. Food and Drug Administration has cleared it as an indirect food additive. Guinea pigs skin sensitized to methyl methacrylate did not react to the polymer.[85] Rats exposed to polymer heated to 100°C for 6 hours showed no toxic effects. Rabbits exposed to polymer dust 15 mg/m³ for up to 30 days showed increased hematoprophyrin levels.[86] It produced solid-state carcinogenic response in animals via implant.[87]
Polymethyl methacrylate 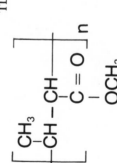 Clear solid or liquid mw, varies, 25,000–1,000,000		
Methyl methacrylate	OSHA–100 ppm TLV–100 ppm	Oral LD50 (rat) is 10 mL/kg.[88] Mild eye and skin irritant, can cause skin sensitization.[89]

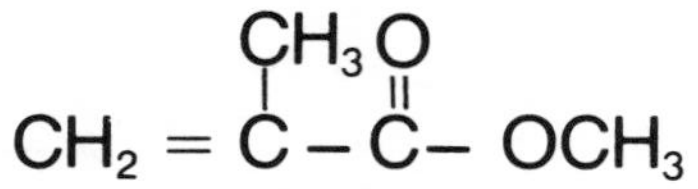

$$CH_2 = \underset{\underset{CH_3}{|}}{C} - \underset{\underset{O}{\|}}{C} - OCH_3$$

Colorless liquid

May be present in small to trace quantities in some uncured resins.

Rats given 2000 ppm in water for 2 years showed temporary weight depression and increased kidney/body weight ratios.[90] In a two-year inhalation study with rats at 400 ppm, no excess cancers were found.[91]

Polyurethanes

OSHA-N.A.
TLV-N.A.

Inhalation LC50 (rat, 4 hr.) is greater than 2460 mg/m³.[75]

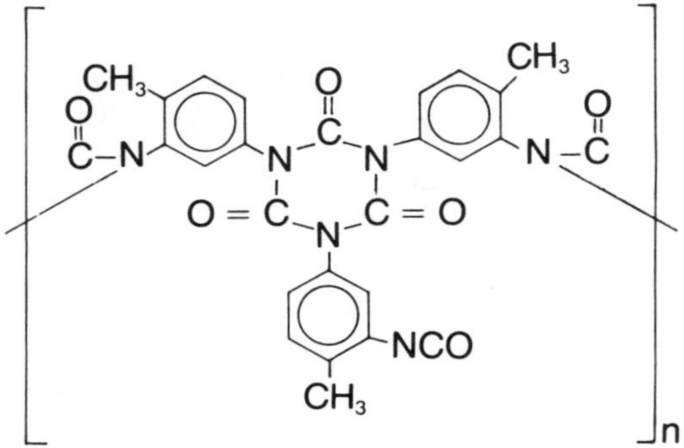

(Table 2 continued)

Toluene diisocyanate May be present in small to trace quantities in uncured resins. CH_3 / NCO / NCO structure		OSHA–0.02 ppm TLV–0.02 ppm (Ceiling)	Oral LD50 (rat) is 5.8 g/kg, skin LD50 (rabbit) is 16 g/kg, LC50 (rat, 4 hr.) is 350 mg/m^3, skin and eye irritant, may cause skin or respiratory tract allergic reactions.[75,92]
$OCN\text{-}(CH_2)_6\text{-}NCO$ Hexamethylene diisocyanate May be present in trace quantities in uncured resins.		OSHA–N.A. TLV–N.A.	Oral LD50 (rat) is 0.71 g/kg, LD50 (rat, 4 hr.) is 310 mg/m^3, skin and eye irritant may cause skin or respiratory tract allergic reactions.[75,92]
Polyvinyl acetate $\left[CH_2 - CH \begin{matrix} \\ O \\ C=O \\ CH_3 \end{matrix} \right]_n$ colorless solid		OSHA–N.A. TLV–N.A.	A copolymer of vinyl chloride/vinyl acetate m.w. 25,000–30,000 fed to rats for two years up to 12% in diet produced no ill effects.[93]

Vinyl acetate

$$CH_3 - \overset{\overset{\displaystyle O}{\|}}{C} - OCH = CH_2$$

Colorless liquid

OSHA–N.A.
TLV–10 ppm

Oral LD50 (rat) is 2900 mg/kg and the inhalation LC50 (rat, 4 hr.) is 4000 ppm.[25] Dermal LC50 (rabbit) 2330 mg/kg. Slight skin and eye irritant. Below 200 ppm rats showed no toxic signs. Human respiratory tract irritation above 20 ppm. A 135-week study with rats at 2500 ppm showed no increase in tumors.[94]

May be present in small to trace quantities in some uncured resins.

Silicone

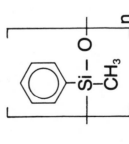

polymethylphenyl siloxane

Viscous liquid

Molecular weight varies

OSHA–N.A.
TLV–N.A.

Oral LD50 (species not specified) is 9 g/kg. Not a skin irritant. Transitory eye irritant. Inhalation exposure of saturated vapors to laboratory animals for 7 hours produced no deaths or visible adverse effects. A one-year rat feeding study, in diet, produced no significant adverse effects. It is cleared by FDA as indirect food additive.[95]

Styrene-butadiene copolymer

OSHA–N.A.
TLV–N.A.

FDA has cleared the copolymer as an indirect food additive.

(Table 2 continued)

$$\left(\!\!\left[CH_2-CH=CH-CH_2\right]_x\!\left[CH_2-CH\right]_y\!\right)_n$$

(phenyl group attached to the $[CH_2-CH]_y$ unit)

Molecular weight range is about 2000–3000. Rosin			
Abietic acid 	OSHA–N.A. TLV–N.A.		Oral toxicity is very low. Not a skin irritant.[96]
Rosin–yellow oil liquid			
DRYING OILS Linseed oil Natural product Oil from flax seed, glycerides of mainly linoleic, linolenic and oleic acids.	OSHA–N.A. TLV–N.A.		Negligible oral toxicity, edible material, cleared as indirect food additive by FDA. It can cause mild skin irritation on human skin.[97] Linoleic acid is generally recognized as safe as nutrient or dietary supplement.

Plastics as a cause of tumor formation have been studied since the 1940's. The types of studies which have received the most attention have been the implant studies. Subcutaneous implants of plastic films in rats produce tumors at the site of implant. Several polymers, including cellulose, as well as metals and glass have produced this phenomenon which has been termed solid-state carcinogenicity or foreign-body tumorigenesis. Plastics (polyethylene) in powder form have not produced tumors when implanted.

From the many studies on the critical properties required for implant tumorigenicity it seems that physical size, shape and surface texture of the implants are responsible for the tumors and not the material's chemical composition. The actual mechanism is not completely elucidated.[15,16]

Phenolics. The first synthetic resins used in paint were phenolic resins which were introduced in the 1920's. Phenolic resins are made from formaldehyde, phenol, or substituted phenols in the presence of alkaline or acid catalysts.

In acid with excess phenol, novolacs are formed. In alkaline solutions with excess formaldehyde, resoles are made. These uncured resins are not encountered in the construction trade.

Vinyl Resins. Vinyl resins were developed after World War II and are polymers of vinyl chloride, vinylidene chloride, vinyl acetate, vinyl alcohol or styrene and its derivatives. These are important resins in paints for the construction trade. Vinyl acetate toxicity is low. Vinyl chloride has produced tumors in laboratory animals and in man and is recognized as a human carcinogen. Resins made from vinyl chloride are manufactured so that only trace amounts (parts per million) of vinyl chloride monomer remain in the polymer.

Alkyd Resins. Alkyd resins are a very important class of synthetic resins and were introduced commercially in the 1930's.

Chemically, this class of material is a condensation product of a polybasic acid and polyhydric alcohol. The poly-functional acid is often phthalic acid and the poly-functional base is glycerol or pentaerythritol. Reaction products are glycerol phthalate or pentaerythritol phthalate. Long chain oils such as linseed, soya, tung and tall oil are often added to give these polyesters specific properties. These products are called oil-modified alkyds. A third type of alkyd is made by reacting alkyd resins with other chemicals such as allyl alcohol, styrene, urea-formaldehyde resins or vinyl chloride to form modified alkyd resins. The toxicity of these resins appears to be low.

Cellulosic Resins. Cellulosic resins are natural product derivatives. They are modifications of cellulose, a linear high molecular weight polymer β-anhydroglucose. Resins in this category are nitrocellulose, cellulose acetate, and cellulose acetate butyrate. They are often used as a coating in the wood furniture industry. Their toxicity is low.

Amino Resins. The two major resins in this category are the urea-formaldehyde resins and melamine-formaldehyde resins which were introduced in the late 1930's and early 1940's, respectively.

These uncured amino resins are used in paints which are cured by baking in the factory. Therefore, only fully cured coating on metals are encountered in the construction industry. They are considered to be low in toxicity.

Chlorinated Rubber. These materials were first produced in the 1930's. Chlorinated rubber is low in toxicity and is treated as an inert dust. The major toxicity concern with these materials is that they may contain about 8 percent carbon tetrachloride which is used as a processing solvent. This retained carbon tetrachloride is not released under ambient conditions, but can be released in the manufacture of end products during heating, mixing or dissolving.

Silicones. Silicone resins are composed of polysiloxanes and were introduced in the 1940's. These materials are low in toxicity.

Methacrylates and Acrylates. Polymers made from acrylic and methacrylic acids and their esters such as methyl methacrylate are popular synthetic resins and were introduced in the early 1950's. They are heavily used in emulsion paints, i.e., waterborne latex paints. The toxicity of the polyacrylates and polymethacrylates (PMMA) is low. PMMA has been cleared as an indirect food additive by the FDA. The monomers are irritating to skin, eyes, and the respiratory tract. The monomers have produced skin sensitization in persons handling the uncured resins.

Styrene-Butadiene Copolymer. Polymers made only from styrene are brittle, but if they are polymerized with butadiene, which itself makes a poor paint binder, a styrene-butadiene copolymer is formed which has good properties as a paint binder. These were the first copolymers used in the paint industry and were introduced in the 1950's. These copolymers are low in toxicity.

Polyisocyanates. Polyisocyanates, commonly called polyurethane resins, are formed by polymerization of a diisocyanate such as toluene diisocyanate (TDI). The reaction products can be polyurethanes, biurets or isocyanurates. These resins have been commerically available since the 1950's. The uncured polyurethane resin contains small amounts of unreacted monomer, and it is the monomer which has been associated with health problems. TDI monomer is relatively low in oral toxicity, but it is a potent skin and respiratory tract sensitizer, even at very low concentrations in air. TDI vapor is extremely irritating to eyes, skin, and the upper respiratory tract. Brush and roller application of polyurethane paints with small amounts of unreacted monomer have not been associated with health problems. However, when these paints are sprayed, proper precautions such as wearing an appropriate respirator should be employed. The OSHA occupational exposure limit is low, 0.02 ppm, and is set to prevent development of respiratory tract sensitization. Fractional ppms of TDI in air can cause a respiratory sensitization response in a sensitized person. Other isocyanate monomers should be considered capable of producing sensitization if exposures are sufficient.

Epoxy Resins. Epoxy resins can be either liquids or solids and are made most often from reactions to bisphenol A and epichlorohydrin. These resins were introduced into commerce in the 1950's. Epoxy resins are not completely cured materials and have been associated with allergic skin reactions and skin and eye irritation. The carcinogenic potential of these resins is unclear; most studies have shown an absence of excess tumors, but a few studies have reported excess tumors. In the latter studies, the resins may have contained excess reactive starting materials.[17] Epichlorohydrin has recently been shown to cause cancer in rats exposed by the inhalation route at 100 and 30 ppm.[18] Because of this new toxicity information, epoxy resins are manufactured to contain little, if any, detectable levels of epichlorohydrin. However, once the resins are cured they do not pose a significant health hazard.

ADDITIVES

Additives are defined as those chemicals which perform a special function or impart a special property to paint. Additives include driers, thickeners, anti-skinning agents, catalysts, plasticizers, surfactants and biocides. Additives are present in the paint at low concentrations, generally in the 0.2 to 10% range. The toxicity of common materials from each group is summarized in Table 3.

The driers are most often metallic salts of lead, calcium, cobalt and manganese. Toxicity is mainly due to metal content.

The most popular thickener for water-based paints is methyl cellulose. Its acute and chronic toxicity is negligible and it has been cleared by the FDA as an indirect food additive.

The plasticizer most often encountered is likely to be one of the phthalates: dibutyl-, diethyl-, diethyl hexyl- and dibutyl sebecate. These are very low in acute oral and dermal toxicity. They are slightly irritating to the nose and upper respiratory tract at high airborne concentrations. Phthalates have recently been shown to produce tumors when fed to laboratory animals.[113]

Benzoyl peroxide is a commonly used catalyst for initiating polymerization reactions. It is low in toxicity by the oral route. It is a mild skin and eye irritant as a dust, but in solution it is less irritating. For some persons it is a skin sensitizer. By inhalation it is moderately toxic. It has not been shown to be mutagenic in bacteria or carcinogenic in animal studies.

Methyl ethyl ketoxime is a common anti-skinning agent. It is low in oral, dermal, and inhalation toxicity. It is an eye irritant.

As a group, the biocides are the most toxic of the additives since it is their function to retard mold and bacterial growth in paint. However, they are present in paint at low concentrations, generally less than 1%, and therefore represent a low human health hazard potential for the applicator.

Phenyl mercuric acetate was perhaps the most common biocide at one time. Its use has diminished as a result of U.S. Government regulations which specify that it can only be used in water-based paints. It is highly toxic due to mercury content. It is a strong skin and eye irritant and has caused allergic contact dermatitis in some individuals. It is not mutagenic or carcinogenic.

Tributyltin chloride or oxide are also biocides. Little toxicity information is presently available.

Tetrachlorophenol represents the chlorinated phenol-type biocide. It is moderately toxic. It is not readily absorbed through the skin. The dust is irritating to eyes, nose, and throat.

Copper salts are also used as biocides. The oxides are used extensively in anti-fouling paint for ships. Systemic toxicity is due to copper content.

Formaldehyde is also used as an effective antimicrobial compound in water-based paints at low concentrations. Formaldehyde is irritating to eyes and upper respiratory tract. At concentrations which are clearly irritating and which can produce tissue damage, above 6 ppm, formaldehyde has produced nasal cancer in rats.[83]

TABLE 3

TOXICITY SUMMARY OF ADDITIVES COMMONLY USED IN PAINTS FOR THE CONSTRUCTION TRADES

Chemical Name Structure Physical State	Occupational Exposure Limit	Toxicity
Benzoyl peroxide	OSHA–5 mg/m³ TLV–5 mg/m	Oral LD50 (rat) is greater than 5000 mg/kg. Inhalation LC50 (mouse) estimated to be 700 ppm. Not a skin irritant (as 78% composite). Irritating to eyes. Skin sensitizer in humans. Negative in modified bacterial mutagen assay.[98,99] Tested as subcutaneous implant in rats for 24 months; no tumors at site.[100] In a mouse skin painting study no significant increase in tumors of treated animals was seen.[101]
White powder Polymerization Initiator		
Cobalt naphthenate	OSHA–N.A. TLV–N.A.	Oral LD50 (rat) is 3900 mg/kg.[102] Mild eye irritant—conjunctivitis, no iris or corneal involvement.[103]
Bluish-red solid		

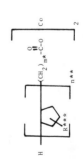

Lead naphthenate

OSHA–N.A.
TLV–N.A.

Oral LD50 (rat) is 5100 mg/kg.[102] Intraperitoneal LD50 (rat) is 520 mg/kg. Oral dosing of 0.25 mL of a 1% solution daily for 4 weeks showed no difference from controls—appearance, weight gain, histology. About 40% of ingested lead was excreted within 3 days.[102]

Yellow solid

Manganese naphthenate

OSHA–N.A.
TLV–N.A.

Oral LD50 (rat) is greater than 6000 mg/kg.[102]

Solid

Driers about 0.5%

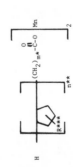

*m > 1
**n = 1–5
***R = small aliphatic groups

Diethyl ethanolamine

$(C_2H_5)_2NCH_2CH_2OH$

Liquid

Curing agent for resins

OSHA–10 ppm
TLV–10 ppm

Oral LD50 (rat) is 1300 mg/kg.[104] Dermal LD50 (rabbit) is 1.26 mL/kg.[105] Severe eye and skin irritant.[106] Respiratory tract irritant. Rats exposed to vapors at 200 ppm 6 hr./day up to 6 months showed weight loss and seven of 50 deaths during first month of study. At end of study no histopathologic abnormalities were observed.[107] Feeding rats 50 to 100 mg/kg/day for 6 months resulted in depressed body weight and increased kidney weight.[107]

Dibutyl phthalate

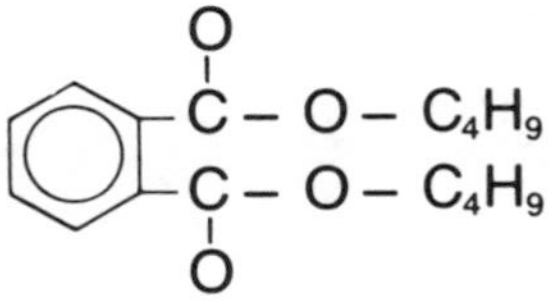

Clear liquid

Plasticizer

5–20%

OSHA–5 mg/m³
TLV–5 mg/m³

Oral LD50 (rat) is 8 g/kg, skin LD50 (rabbit) is greater than 20 mL/kg. Inhalation, cat, at 1 mg/L for 5.5 hr., produced nasal irritation.[108]

(Table 3 continued)

Diethyl phthalate Clear liquid Plasticizer 5–20%	OSHA–N.A. TLV–5 mg/m^3	Oral LD50 (rabbit) is 1000 mg/kg. Slight skin irritant. Subcutaneous LD50 (guinea pig) is 3000 mg/kg. No rats died when exposed to airborne levels of 511 ppm for 6 hours.[108]
Dioctyl phthalate (diethylhexyl phthalate) Light colored liquid Plasticizer 5–20% in paint	OSHA–5 mg/m^{3*} TLV–5 mg/m^{3*}	Oral LD50 (rat) is 30.6 g/kg.[109] Not a skin irritant or sensitizer in human patch test. Chronic feeding studies show no toxic effects at less than 1000 ppm.[110-112] However in a 2-year feeding study at diet levels of 3,000 to 12,000 ppm liver cancer was seen in rats and mice.[113] Inhalation of air saturated with fumes killed all rats exposed for 4 hours, but none of them when exposed for 2 hours.[109]

Di-N-butyl sebecate

OSHA–N.A.
TLV–N.A.

Oral LD50 (rat) is greater than 16 g/kg.[114]

$$C_4H_9O-\overset{O}{\underset{\|}{C}}-(CH_2)_8-\overset{O}{\underset{\|}{C}}-O-C_4H_9$$

Liquid

Plasticizer

5–20% in paint

Formaldehyde
About 1%, biocide

OSHA–3 ppm
TLV–2 ppm
Ceiling

Oral LD50 (rat) is 500 mg/kg. Skin and eye irritant. Eye and upper respiratory tract irritant about 0.5 ppm in humans. Skin sensitizer. It produced nasal tumors in rodents at airborne levels of 6 and 15 ppm.[82,83]

Methyl cellulose

OSHA–N.A.
TLV–N.A.
(cellulose-nuisance dust 10 mg/m³)

Very low in toxicity when administered by normal routes. Cleared by FDA as food additive. Not a skin or eye irritant. Chronic feeding study in animals at 440 mg/kg for 8 months caused no toxic effects. Human exposure to the dust has not produced any known adverse effects.[115]

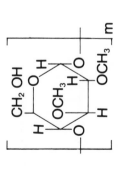

(Table 3 continued)

Variable molecular weight 40,000–180,000.

White granular solid

Thickener for water-based paints.

1–3% in paint.

Methyl ethyl ketoxime

OSHA–N.A.
TLV–N.A.

$$CH_3 - \overset{\displaystyle \overset{NOH}{\|}}{C} - CH_2CH_3$$

Oral ALD (rat) is 3.0 mL/kg•ALC (rat, 4 hr.) is 5000 ppm. Not a skin irritant or sensitizer. Moderate eye irritant.[116]

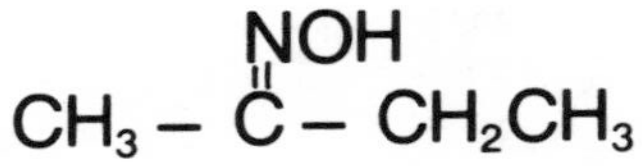

Liquid

Anti-skinning agent

Less than 0.5% in paint

Phenyl mercuric acetate

OSHA–0.1 mg/m³ ceiling
TLV–0.05 mg/m³ (as Hg).

Oral LD50 (rat) is 30–60 mg/kg.[117] Severe skin irritant, can be skin absorbed.[118] Not carcinogenic or mutagenic.[119-121] Toxicity due mainly to mercury content. Mercury poisoning produces a variety of toxic effects on the kidney and central nervous system.[122]

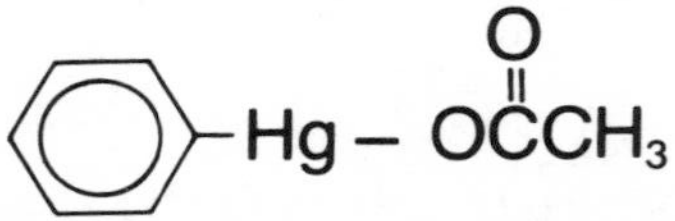

White powder Fungicide 0.5% in paint	OSHA–N.A. TLV–N.A.	Oral LD50 (rat) is 150–465 mg/kg. Not a skin irritant and is poorly skin absorbed. Severe eye irritant. Inhalation of dust can cause eye, nose, and throat irritation, tremors and convulsions.[123,125]
2,3,5,6-tetrachlorophenol Brown solid Fungicide 2% in paint		
Tributyl tin chloride Sn⁺(C₄H₉)₃Cl⁻ Biocide	OSHA–0.1 mg/m³ (as Sn) TLV–0.1 mg/m³ (as Sn)	Systemic toxin which can penetrate skin.[125] Dietary levels of triethyl tin at 10–20 ppm have produced progressive weakness and tremors. Brain damage can occur. Skin burns in workers handling tributyl tin have been reported.[126] Toxicity of tributyl tin is probably between tributyl tin acetate which has an oral LD50 (rat) of approximately 400 mg/kg and bistributyl tin oxide, oral LD50 (rat) is 175 mg/kg.[127]

REFERENCES

1. Lefaux, R. The Toxicology of Macromolecules. *Practical Toxicology of Plastics*, Iliffe Books, Chapter II, 48–60, 1968.
2. Zapp, J. A., Jr. *Arch. Environ. Health 4:*125–136, 1962.
3. Autin, J. Toxicology of Plastics in *Toxicology, the Basic Science of Poisons*, Ed. L. J. Casarett and J. Doull, 604–626, 1975.
4. Eckardt, R. E., Hindin, V. R. *J. Occup. Med. 15:*808–819, 1973.
5. Nylen, P., Sunderland, E. *Modern Surface Coatings*, Interscience, John Wiley, 1965.
6. *Documentation of the Threshold Limit Values*, 4th ed. Am. Conf. Governmental Ind. Hyg., Cincinnati, OH, 364, 1980.
7. *TLVs® Threshold Limit Values for Chemical Substances in Workroom Air.* Adopted by ACGIH for 1981, 32, 1981.
8. Nau, C. A., Neal, J., Stembridge, V.A., Cooly, R. N. *Arch. Environ. Health 4:*415–431, 1962.
9. Von Haam, E., Mallitte, F. S. *Arch. Ind. Hyg. Occup. Med. 6:*237–242, 1952.
10. Ingalls, T. H., Risques-Iribarren, R. *Arch. Environ. Health 2:*429–433, 1961.
11. Polak, L., Turk, J. L., Frey, J. R. *Prog. Allergy 17:*145–226, 1961.
12. *IARC Monographs on the Evaluation of the Carcinogenic Risk of Chemicals to Humans, Some Metals and Metallic Compounds 23:*205–323, 1980.
13. *IARC Monographs on the Evaluation of the Carcinogenic Risk of Chemicals to Humans, Some Aromatic Azo Compounds*, 8, 1975.
14. Milvy, P., Kay, K. *Toxicol. Environ. Pathol. 4:*31–36, 1978.
15. Brand, K. G. Foreign Body-Induced Sarcomas in *Cancer A Comprehensive Treatise 1*, Ed. F. F. Becker, Plenum, 485–508, 1975.
16. Bryson, G., Bischoff, F. *Prog. Exp. Tumor Res. 11:*100–133, 1969.
17. Hine, C.H., Rowe, V. K., White, E. R., Dormer, J., Youngblood, G. T. Epoxy Compounds in *Patty's Industrial Hygiene and Toxicology*, Eds. G. D. Clayton and F. C. Clayton, 3rd rev. ed., 2241-2259, 1981.
18. Laskin, S., Sellakumar, A. R., Kuschner, M., Nelson, N. La Mendola, S., Rusch, G.M., Katz, G. V., Dulak, N. C., Albert, R. E. *J. Natl. Cancer Inst. 65* (4):751–757, 1980.
19. Gross, P., Harley, R., de Treville, R. A. *Arch. Environ. Health 26*(5):227–236, 1973.
20. Gonalewski, G. *Arch. Gewerbepath Gewerbehyg. 11:*108–130, 1941.
21. McLaughlin, A. L. G., Kazantzis, G., King, E., Teare, D., Porter, R. J., Owen, R. *Br. J. Ind. Med. 19:*253–263, 1962.
22. Mitchell, J. *Br. J. Ind. Med. 18:*10–20, 1961.
23. Unpublished data. Du Pont Company, Wilmington, DE.
24. Wagner, J. C. *Br. J. Cancer 28:*173–185, 1973.

25. Smith, H. F., Jr., Carpenter, C. P. *J. Ind. Hyg. Toxicol. 30:*63–68, 1948.
26. Unpublished data. Industrial Bio-Test Laboratories, Northbrook, IL.
27. Unpublished data. PPG Industries, Pittsburgh, PA.
28. Unpublished data. Du Pont Company, Wilmington, DE.
29. Grant, W.M. *Toxicology of the Eye,* 2nd ed., C.C. Thomas, 173-174, 1974.
30. Cember, H., Hatch, T. F., Watson, J. A., Grucci, T. *Arch. Ind. Health 12:*628–634, 1955.
31. Einbrodt, H. J., Wobker, R., Klippel, H. G. *Int. Arch. Arbeitsmed. 30:*237–244, 1972.
32. Hutcheson, D. G., Gray, D. H., Venugopal, B., Luckey, T. D. *J. Nutr. 105:*670–675, 1975.
33. Southam, C. M., Babcock, V. I. *Am. J. Obstet. Gynecol. 96*(1):134–140, 1966.
34. Merck Index, 9th ed., 210, 1976.
35. TLVs® *Threshold Limit Values for Chemical Substances in Workroom Air.* Adopted by ACGIH for 1981, 12, 1981.
36. Unpublished data. Monsanto Co., St. Louis, MO.
37. Nau, C. A., Neal, J., Stembridge, V. A. *Arch. Environ. Health 1:*512–533, 1960.
38. Nau, C. A., Taylor, G. T., Lawrence, C. H. *J. Occup. Med. 18*(11):732–734, 1976.
39. Smyth, H. F., Jr., *Am. Ind. Hyg. Assoc. J. 30:*470–476, 1969.
40. Gilman, J. P. W. *Cancer Res. 22:*158–162, 1962.
41. Pott, F., Huth, F., Friedrichs, K. H. *Environ. Health Perspect. 9:*313–315, 1974.
42. Gilman, J. P. W., Herchen, H. *Acta Unio. Int. Cancrum 19:*615–619, 1963.
43. Saffiotti, V., Montesano, R., Sellakumar, A. R., Cefis, F., Kaufman, D. G. *Cancer Res. 32*(5):1073–1081, 1972.
44. Nettesheim, P., Creasia, D. A., Mitchell, T. J. *J. Natl. Cancer Inst. 55:*159–169, 1975.
45. Jones, J. G., Warner, C. G. *Br. J. Ind. Med. 29:*169–177, 1972.
46. Teculescu, D., Albu, A. *Int. Arch. Arbeitsmed. 31*(2):163–170, 1973.
47. Merck Index, 8th ed., 709, 1968.
48. *Criteria for a Recommended Standard . . . Occupational Exposure to Chromium (VI),* NIOSH #76–129, 1975.
49. Maltoni, C. *Ann. NY. Acad. Sci. 271:*431–443, 1976.
50. Davies, J. M. *J. Oil Colour Chem. Assoc. 62:*159–163, 1979.
51. Iangand, S., Nonseth, T. *Br. J. Ind. Med. 32:*62–65, 1975.
52. Merck Index, 9th ed. 711, 1976.
53. Stockinger, H. E. The Metals in *Patty's Industrial Hygiene and Toxicology,* Eds. G. D. Clayton and F. C. Clayton, 3rd rev. ed., 1687–1728, 1981.

54. Dreessen, W. C., Dallavalle, J. M., Edwards, T. I., Sayers, R. R., Easom, H. F., Trice, M. F. *Public Health Bull. No. 250*, 1940.
55. *Documentation of the Threshold Limit Values*, 4th ed., Am. Conf. Governmental Ind. Hyg., Cincinnati, OH, 362–365, 1980.
56. Tebbens, B. C., Beard, R. R. *Arch. Ind. Health 16:*55, 1957.
57. *Documentation of the Threshold Limit Values*, 4th ed., Am. Conf. Governmental Ind. Hyg., Cincinnati, OH, 381–385, 1980.
58. Unpublished data. Commercial Pigments Corp.
59. Unpublished data. Du Pont Company, Wilmington, DE.
60. Christie, H., Mac Kay, R. J., Fisher, A. M. *Am. Ind. Hyg. Assoc. J. 24:*42–46, 1963.
61. National Cancer Institute: Bioassay of Titanium Dioxide for Possible Carcinogenicity DHEW Pub. No. (NIH) 78–1347, 1978.
62. *Documentation of Threshold Limit Values*, 4th ed., Am. Conf. Governmental Ind. Hyg., Cincinnati, OH, 399–400, 1980.
63. Ibid., 446–447.
64. Ibid, 445.
65. Dalager, N. A., Mason, T. J., Fraumeni, J. F. Jr., Hoover, R. and Payne, W. W. *J. Occup. Med. 22*(1), 25–9, 1980.
66. Unpublished data. Du Pont Company, Wilmington, DE.
67. Haddlow, A., Horning, E. S. *J. Natl. Cancer Inst. 24:*109–147, 1960.
68. Unpublished data. Du Pont Company, Wilmington, DE.
69. Material Safety Data Sheet, Harmon Colors Corp.
70. Unpublished data. Du Pont Company, Wilmington, DE.
71. Back, K. C., Thomas, A. A., MacEwen, J. D. *Natl. Technical Information Service Report PB 225–283*, 1973.
72. Unpublished data. Du Pont Company, Wilmington, DE.
73. Vasilenko, N. M., Zvezdai, V. I., Kolodub, F. A. *Gig. Sanit. 8:*103–4, 1974.
74. Anderson, A. *Brit. J. Ind. Med. 3:*243, 1946.
75. Bunge, W., Ehrlicher, H., Kimmerle, G. *Zentralbl. Arbeitsmed. Arbeitsschutz Prophyl. 4:*20–26, 1977.
76. Material Safety Data Sheet. Chlorinated Rubber. Parlon, Hercules Inc., Wilmington, DE.
77. Anderson, H.H., Hine, C.H., Guzman, R.J., Wellington, J.S. Univ. Calif. Sch. Med., San Francisco, U.C. Report #270, 1957 in NIOSH Current Intelligence Bulletin #29, October 12, 1978, DHEW (NIOSH) Pub. No. 79–104.
78. Gage, J. C. *Brit. J. Ind. Med. 16:*11, 1959.
79. Dilley, J. V., Anderson, B. S., Roberts, D. N., Lee, C. C. *Toxicol. Pharmacol. 33*(1):159, 1975.
80. Unpublished data. Du Pont Company, Wilmington, DE.
81. Diphenyl oxide modified Novaloc, Dow Chemical, Product Information Bulletin, Midland, MI, 1963.
82. Formaldehyde and Other Aldehydes. National Research Council, National Academy Press, 1981.

83. Swenberg, J. A., Kerns, W. D., Mitchell, R. I., Gralla, E. J. and Pavkov, K. L., *Cancer Research 40:*3398–3402, 1980.
84. Unpublished data. Hinkle Corporation, Minneapolis, MN.
85. Nyguist, G. et al. *Odondol. Revy. 23*(2):197–203, 1972 from *CA* 77:168574d.
86. Unpublished data. Du Pont Company, Wilmington, DE.
87. *IARC Monographs on the Evaluation of the Carcinogenic Risk of Chemicals to Humans, Some Monomers, Plastics and Synthetic Elastomers, and Acrolein, 19:*195–211, 1979.
88. *Documentation of Threshold Limit Values*, 4th ed., Am. Conf. Governmental Ind. Hyg., Cincinnati, OH, 285, 1980.
89. Chung, C. W., Giles, A. L. *J. Invest. Dermatol. 69:*187–190, 1977.
90. Borzelleca, J. F., Laison, P. S., Hennigor, G. R., Jr., Huf, E. G., Crawford, E. M., Smith, R. B., Jr. *Toxicol. Appl. Pharmacol. 6:*29–36, 1964.
91. Unpublished data. Rohm & Haas Chemical Co., Philadelphia, PA.
92. Criteria for a Recommended Standard . . . , Occupational Exposure to Diisocyanates, NIOSH #78-215, 1978.
93. Smith, H. F., Jr. and C. S. Weil, *Toxicol. Appl. Pharmacol. 9*(3):501–504, 1966.
94. Criteria for a Recommended Standard . . . , Occupational Exposure to Vinyl Acetate. NIOSH #78–205, 1978.
95. Rowe, V. K., Spencer, H. C., Bass, S. L. *J. Ind. Hyg. Toxicol. 30*(6):332–352, 1948.
96. Fassett, D. W. Organic Acids, Anhydrides, Lactones, Acid Halides and Amides, Thioacids in *Industrial Hygiene and Toxicology*, Ed: F. A. Patty 2:1771–1846, 1963.
97. Frosch, P. J., Kligman, A. M. *Cutaneous Toxicity*. Eds.: V. A. Drill and P. Lazar, 127, Academic Press, 1977.
98. *Criteria for a Recommended Standard . . . , Occupational Exposure to Benzoyl Peroxide*. NIOSH #77–166, 1977.
99. *Documentation of the Threshold Limit Values*, 4th ed., Am. Conf. Governmental Ind. Hyg., Cincinnati, OH, 41–42, 1980.
100. Heuper, W. C. *J. Natl. Cancer Inst. 33:*1005–1027, 1964.
101. VanDuuren, B.L., Nelson, N., Orris, L., Palmes, E.D., Schmitt, F.L. *J. Natl. Cancer Inst. 31:*41–55, 1963.
102. Rockhold, W. T. *Arch. Ind. Health 12:*477–482, 1955.
103. Unpublished data. Du Pont Company, Wilmington, DE.
104. Smith, H. F., Jr., Carpenter, C. P. *J. Ind. Hyg. Toxicol. 26:*269–273, 1944.
105. Smith, H. F., Jr. Unpublished study. Personal communication to TLV Committee in *Documentation of Threshold Limit Values*, 4th ed., Am. Conf. Governmental Ind. Hyg., Cincinnati, OH 140–141, 1980.
106. Safety Data Sheet. ALCOLAC Chem. Corp.
107. Cornish, H. H. *Am. Ind. Hyg. Assoc. J. 26:*479–484, 1965.

108. Fassett, D. W. Esters in *Industrial Hygiene and Toxicology*. Ed: F. A. Patty, 2:1847–1934, 1963.
109. Schaffer, C. B., Carpenter, C. P., Smith, H. F., Jr. *J. Ind. Hyg. Toxicol.* 27:130–135, 1945.
110. Fukuhara, M., Takabatake, E. *J. Toxicol. Sci.* 2:11–23, 1977.
111. Yanagita, T., Kobayashi, K., Enomoto, N. *Biochem. Pharmacol.* 27:2283–2288, 1978.
112. Yanagita, T., Kuzuhara, S., Enomoto, N., Shimada, T., Sugano, M. *Biochem. Pharmacol.* 28:3115–3121, 1979.
113. Carcinogenesis Bioassay of di(2-ethylhexyl)phthalate, NTP-80-37, DHHS #(NIH) 81–1773, 1980.
114. Smith, C. C. *Arch. Ind. Hyg. Occup. Med.* 7:310–318, 1953.
115. Hake, C. L., Rowe, V. K. Ethers in *Industrial Hygiene and Toxicology*. Ed.: F. A. Patty, 2:1711–1712, 1963.
116. Unpublished data. Du Pont Company, Wilmington, DE.
117. *NIOSH Registry of Toxic Effects of Chemical Substances*, 1979 ed., 2: 36, USDHHS, Cincinnati, 1980.
118. Unpublished data. Wells Laboratory.
119. Fitzhugh, O. G., Nelson, A. A., Lang, E. P., Kunzl, F. M. *Arch. Ind. Hyg. Occup. Med.* 2:433–442, 1950.
120. Innes, J. R. M., Ulland, B. M., Valerio, M. G., Petrucelli, O., Fishbein, L., Hart, E. R., Pallotta, A. J. *J. Natl. Cancer Inst.* 42:1101–1114, 1969.
121. Anderson, K. J., Lughty, E. G., Takahaski, M. T. *J. Agric. Food Chem.* 20:649–56, 1972.
122. Stokinger, H. E. The Metals (Excluding Lead) in *Industrial Hygiene and Toxicology*. Ed. F. A. Patty: 2:987–1194, 1963.
123. Antimicrobial Agents IV-6, Dowicide 6, Dow Chemical Co., 1972.
124. Deichman, W. B., Keplinger, M. L. Phenols and Phenolic Compounds in *Industrial Hygiene and Toxicology*. Ed.: F. A. Patty, 2:1403–1404, 1963.
125. *Documentation of the Threshold Limit Values*, 4th ed., Am. Conf. Governmental Ind. Hyg., Cincinnati, OH, 398–399, 1980.
126. Stokinger, H. E. The Metals (Excluding Lead) in *Industrial Hygiene and Toxicology*, Ed: F. A. Patty, 2:1152–1153, 1963.
127. Technical Bulletin. D-112, Carlisle Chemical Co.

CHAPTER 2

SOLVENT TECHNOLOGY IN PRODUCT DEVELOPMENT

Charles M. Hansen

Scandinavian Paint and Printing Ink Research Instititute
Agern Allé 3, 2970 Horsholm, Denmark

INTRODUCTION

The role of solvents in coatings has gradually changed over the years, but recent developments in the air pollution and painter safety sectors are causing major upheavals. In the past the least expensive solvent with satisfactory solvency and appropriate volatility was used. Current choices are governed by these considerations as well as by legislation which restricts the use of certain solvents for the reasons above to certain levels in various countries. Solvent technology has kept up with these increasing demands, and new techniques and new product development will ultimately resolve the current dilemas; compromises will be required as usual, however.

SOLVENT SELECTION, INDUSTRIAL COATINGS

There are many criteria to be met before a solvent composition is finally worked out for a given application. The most common current goal is to arrive at the least expensive blend which meets the requirements of volatility, solvency, and legal restrictions.

About 15 or 20 solvents are used widely in the coatings industry in industrial applications (Table 1). Other candidate liquids have been eliminated largely because of cost, toxicity, odor, or other undesirable properties. The problem of solvent selection then becomes one of determining which combination of solvents is best for the given case. Although older industrial formulations contained as many as 8 or 10 different solvents, most current systems generally contain 4 or less.

Advances in solubility parameter (δ) considerations have aided more rational solvent selection and other technical improvements resulting from improved knowledge of solvent properties (Hansen, 1967, Hansen, 1967, Hansen, 1967, Hansen and Skaarup, 1967, Hansen, 1969, Hansen and Beerbower, 1971, Hansen, 1972, Hansen, 1977, Hansen, 1978, Barton, 1975). In practice, solubility of a polymer is tested in a number of solvents such as those suggested by Figure 1. A circular region of solubility with a radius usually between 3 and 6 units will be defined on this Figure by the good solvents. All possible combinations of solvents capable of dissolving the

43

TABLE 1			
Data for solvents used in coatings			
Solvent	**Evaporation rate**	**Solubility parameters**	
		δ_D $\quad\delta_P$	δ_H

Solvent	Evaporation rate	δ_D	δ_P	δ_H
Acetone	1160	7.58	5.1	3.4
Methyl ethyl ketone	722	7.77	4.4	2.5
Ethyl acetate	615	7.44	2.6	4.5
Methanol	610	7.42	6.0	10.9
Isopropyl acetate	500	7.30	2.2	4.0
Ethanol	340	7.73	4.3	9.5
Isopropanol	300	7.70	3.0	8.0
Toluene	240	8.82	0.7	1.0
Isobutyl acetate	174	7.35	1.8	3.7
Methyl isobutyl ketone	165	7.49	3.0	2.0
2-Nitro propane	125	7.90	5.9	2.0
n-Butyl acetate	100	7.67	1.8	3.1
Isobutanol	80	7.40	2.8	7.8
Xylene	63	8.65	0.5	1.5
n-Butanol	45	7.81	2.8	7.7
Ethylene glycol mono ethyl ether	32	7.49	3.0	2.0
Cyclohexanone	23	8.65	4.1	2.5
Ethylene glycol mono ethyl ether acetate	21	7.78	2.3	5.2
Diacetone	14	7.65	4.0	5.3
Ethylene glycol mono butyl ether	6	7.76	3.1	5.9

polymer can then be located by using a linear mixing rule based on volume fractions of each component.

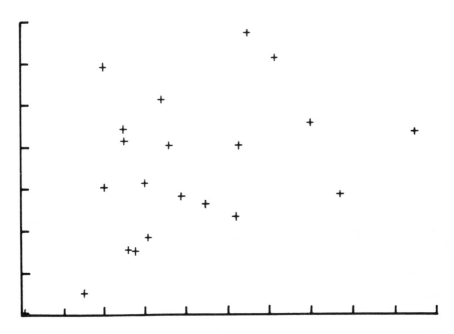

FIGURE 1 Solubility parameter plot for common solvents.

In general the volatility restrictions are sufficient to significantly narrow the number of solvents applicable to a given situation. A common guide is the relative evaporation rate. In one system all volatilities are assigned values relative to n-butyl acetate, which is arbitrarily assigned an evaporation rate of 100. An important part of the volatility question in mixed solvents is solvent balance. The least volatile solvent in the mixture should generally be a good solvent for the polymer. Control of evaporating solvent compositions can be accomplished in a few cases by using azeotropic compositions (Ellis and Goff, 1972). New methods of calculation of residual solvent composition have also been reported (Walsham and Edwards, 1971, Gilbert, 1971, Derr and Deal, 1973, Sletmoe, 1970, Dillon, 1977, Stratta *et al*, 1978, Yoshida, 1972). The principles of solvent evaporation from solution coatings (Hansen, 1968) are different from those of water-borne coatings (Hansen, 1974, Sullivan, 1975). Assuming that the volatility question can be handled, the next restrictions of consequence is compliance with existing legislation.

Water-borne coatings are desirable in this respect but are only slowly gaining acceptance because of poor performance properties. Other approaches include high solids coatings, radiation cured coatings, and powder coatings.

SOLVENT SELECTION, CONSUMER PRODUCTS

The principles of solvent selection are basically the same regardless of whether a coating is intended for industrial or consumer products use. Volatility and solvency are the primary technical considerations while worker safety (legislation) is then dominant within the technical possibilities.

In the building trade about 80% of the coatings now used are water-borne, while various estimates place this figure at 5—10% for industrial coatings. Significant progress has thus been made, but future progress will be slower. The easiest part of the job has been done. The organic solvents still remaining are mostly mineral spirits in solvent-borne products and acceptably small amounts of glycols or ether alcohols in water-borne products. (See Table 2). The ultimate goal is to remove all potentially harmful solvents, of course.

TABLE 2	
Solvents which may appear on the building site:	
Solvent Type[X]	**Found in**
Mineral Spirits (some aromatic content)	Alkyd or oil based products
Glycols, ether alcohols, (amines)	Water borne products
Methylene chloride	Paint Removers
Ketones, esters, alcohols, aromatics	Acid-curing varnishes, vinyl coatings, chlorinated rubber-based coatings.

[X]Two component systems will not be treated in this paper other than to mention that esters, ketones, and aromatics are likely to be present.

PRODUCT PERFORMANCE

In general it can be said that each type of coating has its own special advantages and disadvantages. These make replacement of oil or alkyd based paints difficult, largely by consumer choice.

Special advantages of oil or alkyd based products include good penetration into wood (which makes them necessary for the primer on pressure impregnated wood, for example), good washability and mechanical properties, and the ability to attain high gloss. Certain barrier properties are also an advantage and water-free primers for bare metal seem to be the safest from a general use point of view.

A subjective reason many (most) still prefer solvent borne stains to water borne stains is that their appearance (in the transparent types) is more uniform and more appealing from an esthetic point of view.

Methylene chloride is present in paint removers since it is a quite universal solvent which subsequently evaporates readily. The alternatives such as burning with the evolution of gaseous products, or sanding and scraping, also have their difficulties.

To sum up, product quality demands have helped to maintain an essentially stable but limited market for solvent borne coatings for the building trade. Water-borne coatings are gaining in volume and acceptance because of easy clean-up, low solvent (odor) levels, and improved technical performance. Where technology has not allowed the development of a water-borne replacement, the only recourse is to evaluate the need for suitable personal protection during the given application.

Product liability damage suits can be as difficult to cope with as environmental changes. Both can cost enormous sums of money. The job must be done with a suitable product under suitable working conditions. This is quite clear.

ORGANIC SOLVENTS IN WATER BORNE COATINGS-WORKER SAFETY

Water-borne coatings are currently receiving exceptional interest in both industrial and consumer products (construction site) areas because of their environmental desirability. Lower levels of organic solvents are nevertheless frequently required for these products to function properly. The favored organic solvents for this purpose are the ether alcohols, exemplified by ethylene glycol monobutyl ether and the straight or branched chain alcohols in industrial coatings and low volatility glycols and ether alcohols in consumer products coatings. The latter evaporate very slowly and contribute to retard drying to an acceptable level, since water evaporates too rapidly by itself (evaporation rate 35 at room temperature and 50% relative humidity).

In general the presence of a limited amount of organic solvent contributes to system stability (the polymer is not dissolved, but dispersed), surface tension reduction (the surface tension of water is too high to allow adhesion/application on most substrates since the coating should have a surface tension lower than that connected with the substrate), and ultimate coalescence of the dispersed polymer to a reasonably homogeneous film (Hansen, 1977). Typical solvents which might be considered from a technical viewpoint in these coatings are those located in the marginal solubility region for both water and polymer on a solubility parameter plot. (See Figure 2 and Table 3).

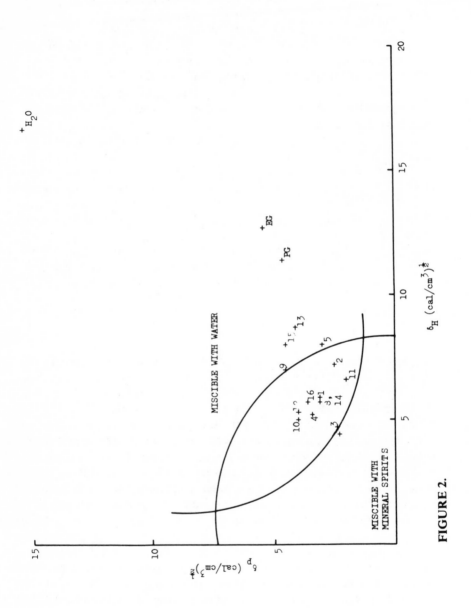

FIGURE 2.

TABLE 3

Volume Fraction of Organic Solvent Required to "couple" Equal Volumes of Mineral Spirits and Water

Solvent	Vol. fraction
1. Ethylene glycol monobutyl ether	54.5
2. *tert*-Butuyl alcohol	60.1
3. Ethylene glycol monoisobutyl ether	68.1
4. Diethylene glycol monobutyl ether	70.2
5. Isopropyl alcohol	71.4
6. Ethylene glycol monoisopropyl ether	76.7
7. Tripropylene glycol monomethyl ether	77.2
8. Dipropylene glycol monomethyl ether	80.8
9. Ethylene glycol monoethyl ether	81.9
10. Tetrahydrofurfuryl alcohol	82.8
11. Cyclohexanol	85.2
12. Diacetone alcohol	85.3
13. Hexylene glycol	85.9
14. Propylene glycol monomethyl ether	89.6
15. Ethylene glycol monomethyl ether	90.4
16. Diethylene glycol monoethyl ether	90.9

The amine present in water-borne coatings is necessary for stability. It associates with carboxyl groups built into the polymer chain making these local regions water soluble where the polymer itself is not (Hansen, 1977). These water soluble regions then migrate to the water phase/polymer phase boundary region and effectively stabilize the dispersed polymer part of the system in the aqueous environment.

Studies on the function of amines in such coatings have shown that at room temperature the amine is essentially totally associated with the acid groups as long as there are free acid groups present, and that the majority of the amine present is retained in the film for very long time (Hansen and Nielsen, 1979).

While there still are organic solvents present in water-borne coatings, these have been able to be reduced to such a low level and most have such low volatilities as to allow these systems to comply with current legislation regarding worker safety. Problems can occur with higher gloss coatings where higher levels of solvent are necessary to effect coalescence to a smooth surface on the film, however. The technical desire that this extra solvent evaporates relatively rapidly to maintain suitable drying time is in conflict with the desire to maintain lowered solvent levels in the air.

DISCUSSION

Those with larger sources of knowledge, expertise, and money are those most likely to make the desired technological breakthroughs in solvent reduction. The formulation of coatings still involves an element of surprise. In spite of advances in science, we don't always know why things work as they do. A given chemist at the laboratory bench in any coatings factory may be the one who makes the big advance, if only he adds the appropriate amount of the appropriate ingredient in the appropriate manner to his already complex formula, the exact interactions of which he only imperfectly perceives. In other words, coatings are complex. Insight into the disciplines of surface chemistry, corrosion, physical chemistry, rheology, polymer chemistry, organic chemistry, materials handling, color science, material science, and environmental science, (and now medical science) among others, are all required at the same time. To this list must be added all general business factors including the cost and availability of raw materials.

I have personally been interested in solvent properties for many years and recognize only too well the relatively limited possibilities to exchange one solvent for another and the problems and time involved when such, even minor changes, are introduced into a well-tried formula.

The substitution of one presumably less harmful solvent for another was the approach of the so-called Rule 66 Changes effected in the United States over the last decade. Industry has shown that it can respond to such demands, but that the approach of legislating arbitrary limits before significant research effort to show their feasibility is not the way things should be done.

Substitute solvents for the optimum solvent based on technical reasons must be present in higher concentrations to achieve the same performance. Substitution of one solvent for another will not necessarily reduce the level of solvent in the air around the person applying the coating. The lower volatility solvents may help do this but drying time may be prolonged beyond acceptable limits and water sensitivity in the final product may result.

Higher solids coatings (lower solvent content) are also being developed, but application problems must be balanced with the desired properties of the final product. A compromise is necessary to make the product acceptable. As with water borne coatings the raw materials going into the coating must be developed to meet all the application and performance requirements. In particular new polymers are required and these must be tried out and often exposed to the environment for prolonged periods of time.

The coatings industry is genuinely grateful for the attention it is now beginning to receive from doctors and other environmentally concerned individuals. This interest in our industries' problems will hopefully stimulate a more vigorous effort to the young minds necessary to develop the technology which will overcome these problems. The coatings industry has accepted the challenge and recognizes the problems involved. Although I have no solution, I am optimistic that progress based on scientific principles and hard work will continue. I also feel the Scandinavian Paint and Printing Ink Research Institute which I represent has contributed significantly to the effort in the past and will continue to do so in the future.

REFERENCES

Barton A F M: Solubility Parameters, Chemical Reviews, 75: No. 6, 731, 1975.

Derr E L and Deal C H: Predicting Compositions During Mixed Solvent Evaporation from Resin Solutions Using the Analytical Solutions of Groups Method in Solvents Theory and Practice, Roy W. Tess (Ed). *Advances in Chemistry Series* No. 124, American Chemical Society, Washington, D.C., 1973, p. 11.

Dillon P W: Application of Critical Relative Humidity. An Evaporation Analog of Azeotropy to the Drying of Water-Borne Coatings, J. Coatings Technology, 49, No. 634, p. 38, 1977.

Ellis W H and Goff P L: Precise Control of Solvent Blend Composition During Evaporation, J. Paint Technology, 44, No. 564, p. 79, 1972.

Gilbert T E: Rate of Evaporation of Liquids into Air, J. Paint Technology, 43, No. 562, p. 93, 1971.

Hansen C M: Doctoral Dissertation, Danish Technical Press, Copenhagen, Denmark, 1967.

Hansen C M: The Three Dimensional Solubility Parameter—Key to Paint Component Affinities I.—J. Paint Technology, 39, No. 505, p. 104, 1967.

Hansen C M: The Three Dimensional Solubility Parameter—Key to Paint Component Affinities II.—J. Paint Technology, 39, No. 511, p. 505, 1967.

Hansen C M and Skaarup K: The Three Dimensional Solubility Parameter— Key to Paint Component Affinities III.—J. Paint Technology, 39, No. 511, p. 511, 1967.

Hansen C M: A Mathematical Description of Film Drying by Solvent Evaporation—J. of the Oil and Colour Chemists Ass., 51, 27, 1968.

Hansen C M: The Universality of the Solubility Parameter—I&EC Product Research and Development, 8, 2, 1969.

Hansen C M and Beerbower A: Solubility Parameters, Encyclopedia of Chemical Technology—2nd Edition. Supplement volume 889, 1971.

Hansen C M: Solvents for Coatings, ChemTech., 2, No. 9, 547, 1972.

Hansen C M: The Air Drying of Latex Coatings. Ind.Eng.Che. Prod.Res. Develop., 13, No. 2, 150, 1974.

Hansen C M: Einige Aspekte der Säure/Base-Wechselwirkung, Farbe und Lack, 83, No. 7, 595, 1977.

Hansen C M: Solvents in Water-Borne Coatings, Ind.Eng.Chem.Prod.Res. Dev., 16, No. 3, 266, 1977.

Hansen C M: Advances in the technology of solvents in coatings., XIV FATIPEC Congress Book (Budapest June 1978) p. 97.

Hansen C M and Nielsen K B: The Behavior of Amines in Water-dilutable Coatings and Printing Inks., J. Coatings Technology, 51, No. 659, 1979.

Sletmoe G M: The Calculation of Mixed Hydrocarbon-Oxygenated Solvent Evaporation J. Paint Tecnology, 42, No. 543, 246, 1970.

Stratta J J, Dillon P W and Semp R H: Evaporation of Organic Cosolvents from Water-Borne Formulations, J. Coatings Technology, 50, No. 647, 39, 1978.

Sullivan D A: Water and Solvent Evaporation from Latex and Latex Paint Paint Films, J. Coatings Technology, 47, No. 610, 60, 1975.

Walsham J G and Edwards G D: A Model of Evaporation from Solvent Blends, J. Paint Technology, 43, No. 554, 64, 1971.

Yoshida T: Solvent Evaporation from Paint Films, Progress in Organic Coatings, 1, 73, 1972.

CHAPTER 3

A COMPARATIVE TOXICOLOGICAL EVALUATION OF PAINT SOLVENTS

M. A. Mehlman, Ph.D. and C. L. Smart, Ph.D.

Mobil Environmental Health and Science Laboratory
P.O. Box 1026, Princeton, N.J. 08540

INTRODUCTION

The solvent-based paints still account for ⅔ to ¾ of the total market consumption in the United States, despite the development of many new paint formulations and a decreased use of organic solvents in paints. Traditional low solids-high solvent paints have been partially but significantly replaced during the past ten years by such technologically newer paints as high solids (70%)-low solvent systems, two-part catalyzed systems, water-based emulsions and latexes, water-soluble and colloidal dispersions, powder coatings, and radiation-cured systems[1] (Figure 1).

The objectives of this paper are to review the paint solvents currently used in the United States, to describe the principal compounds, in, and the amounts of, each class of solvents used, and to summarize the known health effects of each compound on humans and on experimental animals. The accompanying Tables compare the physical properties as well as the toxicological effects of five groups of solvents. These Tables display the information derived from all available acute and long-term studies. Special emphasis is placed on the need for additional toxicological investigations, partly because of the questions being raised by governmental regulatory agencies.

A. ALIPHATIC SOLVENTS

The aliphatic and aromatic solvents rank first in volume as paint and coating solvents. In uses requiring the more expensive oxygenated solvents (alcohols, ketones, and ester/ether), an equal volume of hydrocarbon solvent is usually included. Solvent-based paints always contain some hydrocarbons.[2]

The aliphatics are less effective solvents than the aromatic hydrocarbons, although the latter are more photoreactive. Use of aromatic-free solvents has been increasing in order to comply with such anti-air pollution regulations as Rule 66 of Los Angeles County. Aniline points are relevant to the choice of solvents, since the lower the temperature, the more aromatic-like is the solvent. Table 1 lists the commonly used aliphatic solvents, together with their temperature ranges for distillation and approximate aniline points.

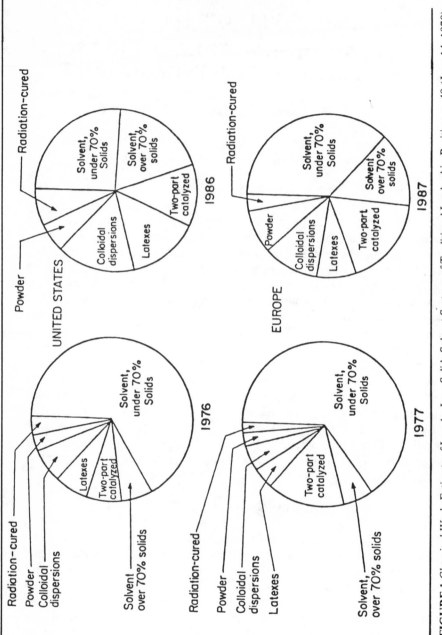

FIGURE 1 *Chemical Week*, Estimate of Loss by Low-Solids Solvent Systems of Traditional Leadership Position, p. 49 (June 14, 1978).

TABLE 1
SELECTED COMPARATIVE FEATURES OF SOLVENTS

	Distillation Ranges, °C	Aniline Point, °C., Approx.	1977 Estimated* Use MM Pounds
A. ALIPHATIC SOLVENTS			
Naphthas	92-210	39-61	
Mineral Spirits	136-277	55-65	
2-Nitropropane	119-122	-	
Kerosene	176-287	54	
n-Hexane	65-70	-	
1-Nitropropane	129-133	-	
Lactol Spirits[1]	203-218	-	
B. AROMATIC SOLVENTS			
Xylene	137-142	9	300
Toluene	110-111	11	255
Ethylbenzene	135-137	10	} 290
Other Alkylated Derivatives[2]	73-282	13-53	
C. ALCOHOL SOLVENTS			
Isopropanol	82-83		125
n-Butyl Alcohol	116-119		151
Isobutyl Alcohol	128-139		200
Methyl Alcohol	63-65		
Diacetone Alcohol	145-175		} 41
Other Alcohols[3]	74-191		
D. KETONE SOLVENTS			
Methyl Isobutyl Ketone	114-117		105
Isophorone	210-218		
Methyl Ethyl Ketone	78-81		357
Cyclohexanone	154-157		
Dimethyl Ketone (Acetone)	55-57		196
2-Heptanone	149-151		
Other Ketone Solvents[4]	104-202		
E. ETHER/ESTER SOLVENTS			
Ethylene Glycol Monoethyl Ether Acetate	150-160		
Isobutyl Acetate	112-119		
n-Butyl Acetate Urene Grade	122-129		86
Ethylene Glycol Monoethyl Ether	134-136		195
Ethylene Glycol Monobutyl Ether	169-173		
Other Ether/Ester Solvents[5]	71-250		

[1] 80% naphtha and 20% kerosene
[2] 100-41-4 (CAS); super hiflash naphtha aromatic hydrocarbon solvent
[3] including 4-methyl-2-pentanol and isodecanol
[4] including 4-methoxy-4-methyl pentanone and N-methyl-pyrrolidone
[5] including m-propyl acetate and 2,2,4-trimethyl-pentanediol-1,3-monoisobutyrate
*Stanford Research Institute

TABLE 2
COMPARATIVE RATINGS OF ACUTE TOXICITY OF SOLVENTS

	Lethality			Irritant Score	
A. ALIPHATIC SOLVENTS	**ORAL**	**DERMAL**	**INHAL**	**EYES**	**SKIN**
Naphthas	1	0	2	1	1
Mineral Spirits	1	1	?	1	1
2-Nitropropane	2	2	2	2	1
Kerosene	0	0	?	1	1
n-Hexane	1	1	?	3	1
1-Nitropropane	2	2	2	2	1
B. AROMATIC SOLVENTS					
Xylene	2	1	1	2	1
Toluene	1	0	1	3	2
Ethylbenzene	2	0	1	2	2
Other Alkylated Derivatives	1	0	2	2	2
C. ALCOHOL SOLVENTS					
Isopropanol	1	0	?	3	0
n-Butyl Alcohol	2	0	?	4	2
Isobutyl Alcohol	2	1	?	4	2
Methyl Alcohol	3	4	?	4	3
Diacetone Alcohol	2	0	?	3	2
Ethyl Alcohol	1	0	1	2	1
D. KETONE SOLVENTS					
Methyl Isobutyl Ketone	2	0	?	2	2
Isophorone	2	1	3	3	2
Methyl Ethyl Ketone	2	0	2	3	2
Cyclohexanone	2	3	?	4	2
Dimethyl Ketone (Acetone)	1	0	0	2	1
2-Heptanone	2	1	?	3	1
Other Ketone Solvents	2	1	?	2	2
E. ETHER/ESTER SOLVENTS					
Ethylene Glycol Monoethyl Ether Acetate	1	0	1	2	2
Isobutyl Acetate	1	0	?	2	1
n-Butyl Acetate Urethane Grade	1	0	?	3	1
Ethylene Glycol Monoethyl Ether	2	1	1	2	2
Ethylene Glycol Monobutyl Ether	2	3	?	3	2

Range of toxicity ratings: 0 (no effect)—4 (highly toxic)

Mineral spirits make up more than 95% of the aliphatic hydrocarbon solvents used in the industry. Their distillation range is 136 to 227°C. The naphthas, used by varnish makers and painters, are also mixtures of aliphatic hydrocarbons but have a lower boiling point and, therefore, evaporate faster than mineral spirits. While well adapted to spraying applications, the fast-evaporating naphthas are generally unsuitable for interior finishes.

Before discussing the toxicological aspects of aliphatic and other solvents, it is helpful to recall the types of testing used to determine potential hazards. The acute effects of paint solvents on humans have been known for some time.[3,4] They include general depression of the central nervous system, irritation of the skin, eye, nose, throat, and lungs, gastrointestinal disturbances, and damage to such visceral organs as the liver and kidney.

Worker exposure during the manufacture of solvents or of solvent-containing paints is less than during actual painting. The routes of exposure in either case are by inhalation, ingestion, eye or skin contact, or a combination of these. The toxic effects depend on the dose—i.e., the duration of exposure and the concentration of the vapor or liquid. Physical exhertion, by augmenting respiratory minute volume, increases the inhalation of vapor.

Acute tests on laboratory animals have been conducted for most of the solvents. The results of these tests indicate the severity of the reaction that may occur on contact with eyes, skin, lungs, and other mucosal areas. These tests are performed on laboratory rodents and rabbits, with the substance either applied to the skin or eyes, fed to the animal, or inhaled by the animals in a vapor-inhalation chamber.

Little work has been done on the chronic effects of these substances on animals. This would require long-term study over the lifetime of an animal and involve oral feeding or gavage, skin painting, and inhalation. Evaluation of the results would determine physiopathalogical effects such as tumor formations, reproduction and gestation, deformed offspring, genetic changes, and other pathological changes should they occur from chronic exposure.

The aliphatic solvents are somewhat irritating to the skin and eyes of laboratory animals. Table 2 shows the acute toxicity ratings of solvents, with the score "0" representing no effect and "4" indicating a highly toxic agent. The aliphatics receive scores of 1 to 3. The aliphatic solvent most irritating to the eye is n-hexane, while all aliphatics have about the same level of irritation on the skin. The two most lethal aliphatic solvents, 2-nitropropane and 1-nitropropane, act through the oral and dermal routes. The inhalation lethality scores are known for naphthas and the nitropropanes, but unknown for mineral spirits, kerosene, and n-hexane.

An excellent solvent for acrylics, epoxies, and cellulosics, 2-nitropropane has been demonstrated to induce liver tumors in one strain of rats, but epidemiological studies do not show adverse health effects on humans exposed to the solvent for 20 years.[5] It is often impossible to predict the potential effects of a substance on humans based on the results of animal studies. Kerosene, n-hexane, and 1-nitropropane are now being studied for chronic effects, but the results may not be available for several years.[6,7]

B. AROMATIC SOLVENTS

With the exception of Hi Flash Solvent Naphtha (Footnote 2 of Table 1), the aromatic solvents are almost pure compounds. They do not require the oxygen-containing solvents, and are used extensively for alkyds and other resins that are insoluble in aliphatics. In addition, the aromatics are used in mixed thinners for lacquers and other coatings that require such oxygenated solvents as alcohols, esters, ethers, and ketones. The aromatic commercial xylene (xylol) is a mixture of the ortho, meta, and para isomers. Xylene, alone or mixed with toluene, is widely used in lacquers and enamels that are applied by spraying. Toluene is used in the same types of coating as is xylene, but it evaporates more quickly.

The aliphatic solvents show a poor correlation between oral and dermal lethality in animals. For example, ethylbenzene has an oral lethality rating of 2 and a dermal lethality of 0, suggesting poor dermal absorption. As a group, the aromatic solvents are more irritating to the eye and skin than are the aliphatic solvents (Table 1).

There is no uniform occupational exposure standard for xylene, toluene, and ethylbenzene: For example, ACGIH and NIOSH recommend 100 ppm for toluene, while OSHA recommends 200. OSHA sets a TLV of 100 ppm for xylene and ethylbenzene, but the Environmental Protection Agency has recently identified these substances[8] as candidates for in-depth risk assessment studies, which could result in new regulations. The selection was based on the large volume of use, high exposure potential, known toxicity, suspected carcinogenicity, and lack of necessary data. Chronic studies[9,10,11,12] are underway on xylene and on toluene, while ethylbenzene has been tentatively selected by the National Cancer Institute for carcinogenic bioassay.[9,13]

C. ALCOHOL SOLVENTS

The alcohol solvents are usually water-soluble and are preferable to the hydrocarbons for forming polar films (shellac, cellulosics, vinyls, acrylics, epoxies, polyurethanes, and silicones). The alcohols are often used as lacquer solvents. As listed in Table 1, the three widely used alcohols are isopropyl, n-butyl, and ethyl. Isopropyl[14] and ethyl alcohols are used in nitrocellulose

Table 3 lists the occupational exposure standards for the aliphatic solvents. They are arranged in order of decreasing OSHA Time Weighted Average (TWA), starting with naphthas and mineral spirits at 500 ppm, and ending with

2-nitropropane and 1-nitropropane at 25 ppm. There are no enforced standards for kerosene and lactol spirits. The OSHA standards for the nitropropanes are the same as those recommended by the American Conference of Governmental Industrial Hygienists (ACGIH), which has also recommended a lowering of the TWA for n-hexane to 50 ppm.

TABLE 3
ALIPHATIC SOLVENTS

OCCUPATIONAL EXPOSURE LIMITS, ppm, TWA

SOLVENTS	OSHA**	NIOSH RECOMM.***	ACGIH[a]
Naphthas	500	75	-
Mineral Spirits	500	55	-
2-Nitropropane	25	treat as a human carcinogen	25
Kerosene	-	14	-
n-Hexane	-	100	50
1-Nitropropane	25	-	25
Lactol Spirits	-	-	-

**up to 8 hour exposure.
***up to 10 hour exposure.
[a]TLV/TWA, 1980 modification. American Conference of Governmental Industrial Hygienists.

The NIOSH Criteria Document on refined petroleum solvents recommends lower standards for naphthas (75 ppm), and mineral spirits (55 ppm) than those proposed by OSHA. In addition, the Criteria Document recommends a standard for kerosene (14 ppm) and regards 2-nitropropane as a potential carcinogen.

There is considerable disagreement among scientists and regulatory personnel over the recommendations contained in the Criteria Document. Additional studies are clearly needed; sufficient data are simply not available. For instance, chronic effects are not known for the naphthas and mineral spirits. Acute symptoms in laboratory animals, which include central nervous system effects and irritation to eyes, nose and throat, are not necessarily relevant to chronic effects. Moreover, these solvents contain benzene, which in high doses has been demonstrated to produce blood abnormalities, and is itself a suspected leukemogen. Thus, chronic testing must take into account, among other things, the concentration of benzene in each solvent.

lacquers and shellac; n-butyl alcohol is used in nitrocellulose coatings. Methyl alcohol[15] is an effective solvent, but its rapid evaporation limits its use.

Most alcohol solvents are highly irritating to the eyes (Table 2). The systemic toxicity of methyl alcohol is well known, causing blindness and death if ingested, inhaled, or absorbed through skin. The EPA has also identified n-butyl and isobutyl alcohols as potential high-risk solvents, subject to possible regulation. Little has been done to determine their long-term effects on animals or humans. Isopropyl alcohol showed no tumorigenic effects in mouse inhalation studies lasting 8 months.

A rat carcinogenesis study is now underway in England in which isopropyl alcohol is administered in the drinking water.[13] A previous reproduction study indicated retarded growth of rat progeny after feeding the mother 2.5% of this alcohol in the drinking water. Several mutagenicity studies, with negative findings, have been carried out with n-butyl and ethyl alcohol.

The recommended occupational exposure standard for the alcohol solvents range from 50 to 1,000 ppm. The most potentially hazardous is isobutyl alcohol, with a TWA of 50 ppm; the least hazardous alcohol solvent is ethyl, with a TWA of 1,000 ppm.

D. KETONE SOLVENTS

Among the solvents containing the oxygen atom, the ketones are more widely used than alcohols, esters, and ethers. The three most frequently used ketones are methyl ethyl, ethyl isobutyl, and acetone. Resins such as unmodified polyurethanes, epoxies, acrylics, vinyl copolymers, and certain cellulosics are insoluble in hydrocarbons or alcohols; and, therefore, require ketone solvents.

Most of the commonly used ketones apparently cause organ damage at high concentrations.[16] Some are now being studied for chronic effects. Cyclo-hexanone and 2-heptanone are being investigated for mutagenic effects, and isophorone for tumorigenic effects. Methyl ethyl ketone showed embryotoxic, fetotoxic, and possible teratogenic effects in rats exposed to 1,000 and 3,000 ppm.[17] The EPA has placed cyclohexanone and methyl isobutyl ketone on the potentially high-risk solvents list. In the past, peripheral neuropathy was attributed to methyl isobutyl ketone exposure, but it has been found that the purified compound does not cause this condition.[18]

The acute lethality and irritation ratings for the ketones as a group are about the same as for the alcohols. Cyclohexanone has the highest level of toxicity in acute testing, followed by isophorone. However, for occupational exposure standards, the TLV for isophorone is lower than that of cyclohexanone. The lack of correspondence between results of acute tests and the working standards is explained by the chronic effects of isophorone, which occur at a lower concentration than those of cyclohexanone.

E. ETHER/ESTER SOLVENTS

Approximately 450 MM annual pounds of ether (derivatives of ethylene glycol) and esters (usually acetates), and ether/esters solvents are used in the United States. These compounds are characterized by high boiling points, high cost, and excellent solvent action. the ethers and esters are somewhat irritating to animals in acute toxicity studies (Table 2). The usual acute effects are well documented, but chronic testing studies have not been completed. The NCI is currently testing ethylene glycol monoethyl ether for carcinogenicity. The EPA has identified it and ethylene glycol monobutyl ether as first priority substances for risk assessment studies. In early 1981, producers of monomethyl ether and monoethyl ether reported to EPA, under the requirements of the Toxic Substances Control Act, Section 8(e), finding testicular changes in laboratory animals inhaling the former, and teratogenic effects in laboratory animals inhaling the latter.

F. CONCLUSION

The occupational exposure standards of paint solvents are being re-examined. The primary concern for this re-examination is potential carcinogenicity. Only 2-nitropropane has been identified as an animal carcinogen, but the benzene contained in naphthas, mineral spirits, and kerosene has raised the question of the potential carcinogenicity of these aliphatic solvents. Carcinogenic bioassays for 1-nitropropane, xylene, ethylbenzene, isopropanol, n-butyl alcohol, and ethylene glycol monoethyl ether are in progress. Depending on the results, either additional solvents will be tested, or extensive testing will cease. Epidemiologic studies have not defined the carcinogenicity among workers exposed to solvents in general. Most studies show irritation of the mucous membranes of the ocular, dermal, respiratory, and gastrointestinal systems. There are also instances of poisoning, characterized by depression of the central nervous system and peripheral neuropathy.[19]

Studies on mutagenicity, reproduction, and teratology are in progress for many paint solvents, but no chronic toxicological information is available for most such solvents. The Appendix summarizes the known information for solvents, as well as the studies in progress. It is apparent that as long as these solvents continue to be used in paints, more animal and human epidemiological studies will have to be initiated to determine their potential for human damage. Without this knowledge, the setting of permissible levels is at best an educated guess. At most, it is an unfounded reassurance.

REFERENCES

1. *Chemical Week*, p. 49 (June 14, 1978).
2. Stanford Research Institute Estimates.
3. See also Draft (subject to revision) of Criteria for a Recommended Standard . . . Occupational Exposure in the Manufacture of Paint and Allied Coating Products, NIOSH.
4. Criteria for a Recommended Standard . . . Occupational Exposure to Refined Petroleum Solvents, NIOSH Publication No. 77-192, p. 132 (1977).
5. Cover letter and booklet—"Review of Safety Data on 2-Nitropropane (2-NP)" from O. W. Chandler, International Minerals and Chemical Corporation, (June 22, 1979).
6. *TOX-TIPS*, National Library of Medicine, Toxicology Information Program.
7. Criteria for a Recommended Standard . . . Occupational Exposure to Alkanes (C5-C8), NIOSH Publication No. 77-151.
8. *Current Reports*, The Bureau of National Affairs, p. 857 (August 24, 1979).
9. National Cancer Institute, Carcinogenesis Testing Program, Chemicals on Standard Protocol, July 2, 1979.
10. Criteria for a Recommended Standard . . . Occupational Exposure to Xylene, NIOSH Publication No. 75-168.
11. *Toxicology, 12*, No. 2, 111 (1979).
12. *Federal Register, 44*, No. 143 (July 24, 1979).
13. Information Bulletin on the Survey of Chemicals Being Tested for Carcinogenicity, No. 7, WHO, International Agency for Research on Cancer, Lyons, France (1978).
14. Criteria for a Recommended Standard . . . Occupational Exposure to Isopropyl Alcohol, NIOSH Publication No. 76-142 (1976).
15. Criteria for a Recommended Standard . . . Occupational Exposure to Methyl Alcohol, NIOSH Publication No. 76-148 (1976).
16. Criteria for a Recommended Standard . . . Occupational Exposure to Ketones, NIOSH Publication No. 78-173.
17. Schivetz, B. A., et al., *Toxicol. Appl. Pharmacol.*, 28, 452 (1974).
18. Spencer, P. S. and Schaumburg, H. H., *Toxicol. Appl. Pharmacol.*, 37 (1976).
19. Registry of Toxic Effects of Chemical Substances, Quarterly Issue, NIOSH (April, 1979); see also Current Intelligence Bulletin Reprints, Bulletins 1 through 18, NIOSH (1978).

APPENDIX A—ALIPHATIC SOLVENTS

Solvent	Carcinogenicity, Mutagenicity Teratology, Reproduction	Health Effects
Naphthas	Unknown, but contains benzene.	**Human:** CNS effects; dermal, eye, nose, throat irritation. **Rats and dogs:** 1200 ppm for 6 hr. day, 5 day week for 65 days, no untoward effects.
Mineral Spirits	Unknown, but contains benzene.	**Human:** Skin irritation. Lung irritation in *lab animals* exposed to 212 ppm for 8 hr. day, 5 day week for 6 weeks.
2-Nitropropane	Mutagenic screening by inhalation exposure underway. Liver cancer in certain male rats.	**Human:** skin, eye, upper respiratory irritant. Epidemiology: no unusual patterns in production workers since 1955.
Kerosene	Unknown, but contains benzene; inhalation route for teratongenic examination underway.	**Human:** skin irritation, especially if high in % aromatics.
n-Hexane	Inhalation, rat mutagenesis and rat teratogenic-potential studies underway. Selected for carcinogenicity testing.	**Human:** polyneuropathy (diseased nerves; causes tingling, numbness, weakness, burning pain).
1-Nitropropane	Mutagenicity - Ames-negative.	**Human:** exposed found concentrations > 100 ppm irritating.

APPENDIX B—AROMATIC SOLVENTS

Solvent	Carcinogenicity, Mutagenicity Teratology, Reproduction	Health Effects
Xylene	Carcinogenicity testing by NCI underway. NIOSH concluded that xylenes, if benzene-free, are not myelotoxic. Oral administration to rat caused reduced fetal weight, as well as fetal formations and maternal toxicity.	**Human:** CNS depressant; airway irritant.
Toluene	Inhalation exposure of rats <300 ppm, 6 hr/day, 5 days/ week for two years: no pathological changes. At high doses causes teratogenicity in developing chick embryos. Carcinogenesis bioassay to start in 1981.	**Human:** CNS depressant; reversible liver injury on long exposure. Human subject study underway on interaction of toluene, methyl ethyl ketone and xylene.
Ethylbenzene	Tentatively selected for testing by NCI.	**Human:** CNS effects; eye, nose, throat, skin irritant. NIOSH plans to complete a criteria document.

APPENDIX C—ALCOHOL SOLVENTS

Solvent	Carcinogenicity, Mutagenicity Teratology, Reproduction	Health Effects
Isopropanol	2.5% in drinking water of rats, retarded growth of first generation progeny. Carcinogenesis study with rat (to be published). Inhalation studies of mice for 4 to 8 months demonstrated isopropanol is non-tumorgenic.	**Human:** mucous membrane irritation.
n-Butyl Alcohol	Non-mutagenic in Ames Test; tentatively selected for testing by NCI.	**Human:** primary skin irritant; eye, nose, throat irritant; dizziness.
Isobutyl Alcohol	Unknown	
Methyl Alcohol	Inadequate experimental animal studies.	**Human:** mild dermatitis; severe poisoning from skin absorption; visual disturbances; metabolic acidosis; death.
Diacentone Alcohol	Unknown	**Human:** dermatitis; eye, nose, throat irritant; CNS effects.
Ethyl Alcohol	Non-mutagenic in Ames Test.	**Human:** dermatitis; eye, nose irritation; headache; CNS effects.

APPENDIX D—KETONE SOLVENTS

Solvent	Carcinogenicity, Mutagenicity Teratology, Reproduction	Health Effects
Methyl Isobutyl Ketone	Purified compound does not cause, as once reported, peripheral neuropathy in animals.	**Human:** eye, nose irritation; CNS effects; gastrointestinal disturbances.
Isophorone	Bioassay begun, testing now complete (gavage, rat and mouse). Results not announced.	**Human:** fatigue and malaise. **Lab animals:** no effects for 8 hr. day, 5 day week, 6 weeks, at 25 ppm; liver and kidney damage at 50 ppm.
Methyl Ethyl Ketone	Embryotoxic, fetotoxic, and potentially teratogenic for pregnant rats exposed to 1,000 or 3,000 ppm, 7 hr. day.	**Human:** eye, nose, throat irritation; human subject study underway of interaction of toluene, xylene and methyl ethyl ketone.
Cyclohexanone	Mutagenic screening by inhalation exposure underway.	**Human:** liver, kidney damage.
Dimethyl Ketone (Acetone)	Found no report implicating ketones as carcinogens or mutagens. All common aliphatic/alicyclic ketones except methyl isoamyl reported to cause organ damage.	**Human:** exposure at 1,000 ppm 3 hr. day, for 7 to 15 yrs. resulted in inflammation of respiratory tract, stomach, duodenum; eye irritant.
2-Heptanone	Advanced mutagenic screening by inhalation exposure underway.	**Human:** mucous membrane irritation.

APPENDIX E—ETHER/ESTER SOLVENTS

Solvent	Carcinogenicity, Mutagenicity Teratology, Reproduction	Health Effects
Ethylene Glycol Mono-ethyl Ether Acetate	Unknown	**Human:** acetate derivative is more irritating than parent compound to the eye.
Isobutyl	Unknown	**Human:** anesthesia and primary irritation characterize most simple esters.
n-Butyl Acetate Urethane Grade	Unknown	**Human:** eye, skin, mucous membrane irritation; CNS effects.
Ethylene Glycol Monethyl Ether	Currently being tested for carcinogenicity in rat and mouse. Completed, but results not announced.	**Human:** mild skin irritant; nose, throat irritant; CNS, liver, kidney effects; pulmonary edema.
Ethylene Glycol Monobutyl Ether	Unknown	**Human:** liver, kidney effects; pulmonary edema.

CHAPTER 4

THE INFLUENCE OF ENVIRONMENTAL FACTORS ON RESEARCH AND DEVELOPMENT OF PAINT PRODUCTS

Lennart Dufva
Manager, Research and Development
AB Wilh, Becker, Trade Paint Division
11783 Stockholm, Sweden

Everyone is discussing a better working environment and less hazardous products nowadays. But in the Sixties we knew scarcely anything about these matters. When we referred to a paint using water as the solvent, its technical advantages were stressed, not the absence of organic solvents. A better working environment was never a sales argument.

In the early Seventies, however, new slogans arose. We began to hear about protecting the environment, pollution control, and less hazardous products. This set off a general discussion about many products potentially damaging to people and the environment. We queried the products used by professional painters and focused particularly on the organic solvents that had long been used in many paints.

Whitewashes and distempers which were used before 1950 were thinned with water and contained no organic solvents. Oil paints contained very little solvent, usually about ten percent. For many years, wood turpentine was used, until gradually replaced by white spirit.

In the Sixties, requirements for interior decoration became greater. This applied not only to houses, but also to offices, schools, hospitals, factories, and so on. It became impossible to use whitewash, distemper, or oil paints. The spread of new building methods outdated the use of oil paints for walls and ceilings. Some of the new building materials were strongly alkaline, like concrete, and saponified the oils and damaged the paints. In addition, demands for faster drying paints with longer life and more resistance to yellowing arose in the market.

To meet these demands, oil paints were replaced by alkyd paints. They have a polymer binder that mainly consists of natural air-drying oils like linseed oil or tall oil. However, the oil molecules were enlarged by letting them react with phthalic anhydride. The new binders dried faster, yellowed only slightly, and resisted chemicals better. The big molecules required more organic solvent, however, than the small molecules of the natural oils. Oil paints had needed about ten percent of the white spirit solvent, whereas the new alkyd paints required about fifty percent. In other words, applying one litre of alkyd paint generated four to five times as much solvent vapor as one litre of oil paint.

The building rush of the Sixties causes a higher rate of paint use and application in alkyd paints' predominance. Rollers became common, and on large surfaces spraying was used.

In addition to the alkyd paints, other types of paint were introduced, e.g., epoxy paints for floors and other surfaces with severe requirements. These special paints needed solvents like alcohols, esters and aromatic compounds.

All these factors meant higher concentrations of organic solvents in the air that painters had to breathe. This meant *more discomfort* and *greater health hazards*.

We can understand the problems of health hazards that arose 10 to 20 years ago. Remember that most of the ceilings, floors, woodwork were being treated with alkyd paints containing solvents, usually white spirit, as well as big stretches of walls. However, many other walls were being treated with latex paints that could be thinned with water. Latex paints had been launched as far back as 1952. Although the manufacturers had succeeded in using water to replace the organic solvents, this was *not* the main reason for developing latex paints. The *main reason* was to achieve technical advantages like faster drying. A latex paint dries in about one hour, while alkyd paint needs about eight yours and an oil paint need at least 24 hours.

Enormous quantities of paints with organic solvents were used in buildings before people were aware of the hazards of inhaling the vapor of organic solvents. However, manufacturers were very well aware of the discomfort the vapor caused as well as the health hazards due to various other components of paints, and measures were already being taken to lessen this discomfort.

As far back as 1955, the initiative was taken by the Swedish Paint and Printing Ink Manufacturers Association. They promoted the Joint Committee for the Occupational Health of Painters, called the *YSAM Committee*. It included representatives of Swedish paint manufacturers and had people from the Swedish Federation of Painting Contractors and from the Swedish Painters' Union. The YSAM Committee cooperated with the State authorities.

One result of the committee's work is a system of marking products, called the YSAM system. It came into force in 1962 and aimed to provide extra information beyond the legal requirements. This made it easier for a painter to select the proper health precautions.

Letter	Production Composition	Precautions
A	Solvent is water	None in premises with normal ventilation
B	Solvent is white spirit or alcohols (except butanol)	None in premises with normal ventilation
C	Solvents not covered by B	Good ventilation required
D	Special regulations apply, e.g., because of legal requirements	special precautions necessary, specified on each package, e.g., lead pigments

Note that the code only applies to products applied by roller or brush, *not* to products applied by spraying.

If we examine the system from today's viewpoint, we note that white spirit was treated as a very harmless solvent. No precautions were laid down for painting rooms with products containing white spirit. In other words, a latex paint and an alkyd paint were considered equivalent with respect to health precautions. No extra ventilation was considered necessary unless paints contained "strong" solvents like butanol, xylene, etc.

One very noticeable difference between an alkyd paint and a water-reducible latex paint, however, was the strong smell of the alkyd paint. This became a good reason for selecting a latex paint over an alkyd paint. For paint manufacturers, this naturally provided an incentive to improve latex paints that could be thinned with water, and to provide a wider range of them.

By 1965 latex paints for walls had become well accepted. Every major manufacturer offered a complete range of grades, shades, and types of gloss. Between 1965 and 1970, latex paints developed greatly. In the early types, the principal binder had been polyvinyl acetate. Polyacrylates now made their entry. They adhered satisfactorily to old coatings of alkyd paints and oil paints, which meant latex paints could now be used for repairs and repainting jobs.

The technical advantages of latex paints deserved greater use by painters, but the new latex paints met with some scepticism. It is quite normal to find that new type of paint only gradually displace older types, even if the new types are much better from the point of view of health hazards. Naturally, a number of reasons account for this: 1) new products and rarely fully developed when they are launched for the first time; 2) new techniques of application may be required; 3) the job specifications may designate the old type of paint, simply because the new one did not exist when they were drawn up; 4) lack of knowledge may exist about the properties of the new products; and 5) the new products often cost more than the old ones. So it takes time for a new product to replace one that is old and well-tried.

The development of latex paints has been greatly stimulated by the growing interest in environmental matters. The public debate has helped, and so have the initiatives of local groups. We can quote the example of paints for Swedish hospitals. Previous specifications required alkyd paints to be used in premises such as rooms for patients, operating theaters, rooms for washing utensils, etc. Latex paints were not considered capable of meeting the health requirements.

However, the County Council of North Bothnia took the initiative and invited some paint manufacturers to make tests. Hospital rooms were to be painted with latex paints only. Both the work of applying the paint and the final results proved that the latex paints were completely satisfactory. In spite of the good results, it took a long time for latex paints to win general acceptance. Even today, you can find doubting Thomases.

I have already listed some of the reasons why new types of paint met with difficulty in displacing old ones. Yet another obstacle is sometimes met with the standard laboratory tests that are carried out to see if a paint is good enough for a particular purpose. We may quote an example from Denmark.

A test for checking the adhesion of a system of painting was used for gypsum wall boards. The coat of paint must adhere so firmly that it does not come off when a strip of adhesive tape is suddenly pulled off. An alkyd system of traditional type survives this test, but a latex system does not. The result is that gypsum boards in Denmark, where this test was employed, must be primed with alkyd paints. In Sweden, however, the tape test was not used. I personally have never heard of practical requirements for better adhesion than a 100 percent latex system provides.

Bit by bit, awareness of the health hazards of organic solvents has spread. The YSAM system was modified in order to ensure better instructions about health precautions. A new code was worked out in 1972. The new system was confined to warning of the risks of breathing the vapor of organic solvents during the application of paints by brush or roller. Other volatile components of paint were not mentioned, and all other health hazards were stated according to official Swedish regulations.

The new system was based on current data on occupational health. These were the threshold limit values laid down by the National Swedish Board of Occupational Safety and Health. Latex paints were grouped under the code marking O, with no requirements laid down for protection of the lungs. Most of the alkyd paints were placed in group 1: for small areas of woodwork, no protection was necessary, but for large surfaces like walls and ceilings, a face mask with gas filter was necessary.

The practice of using alkyd paints for walls began to dwindle until December 1974, when it ceased altogether. This followed a decision of the Swedish Association of Paint Manufacturers which ended all sales of this type of paint.

At the same time, another decision was passed to stop using alkyd paints for ceilings if possible. This type of paint is still being used, however, on a small scale. So far, it has not been possible to make a latex paint suitable for repairing ceilings damaged by water or stained by nicotine.

As I said earlier, the YSAM system laid down no health precautions for the use of alkyd paints on woodwork. This was probably one reason why launching water-reducible latex paints for woodwork had limited success. In 1969, primers were launched, and in 1970, latex-based finishing paint was on the market. Certainly these latex paints had a number of technical disadvantages, which outweighed the advantages of being able to use water as the solvent instead of white spirit. These disadvantages referred to flow and gloss, but perhaps more importantly the techniques of application. New techniques were need to ensure good final results because of the special properties of latex paints. Some painters soon learned how to master the new paints and began using them at an early date. But the majority kept using alkyd paints of traditional type, in spite of the organic solvents.

However, in 1978–1979, there has been a marked increase in the use of water-reducible woodwork paints. Primers of latex type have been firmly established for several years. Present consumption is at least as great as that of alkyd primers. However, the use of latex finishing paints has not yet come as far.

One reason for the great increase in the interest in water-reducible woodwork paints is the fact that the health hazards of white spirit are regarded more seriously. Another reason is that almost all other kinds of interior painting and decorating are done with water-reducible paints. The result is that the difference in paint smell has now become even more evident. The change-over will probably take place fairly fast now that the market has begun to offer new, improved woodwork paints that are water-reducible.

What is the present situation of paints for the interiors of buildings? Almost all types of work can now be done with water-reducible paints. This is particularly the case for parquet floors. A parquet lacquer (latex type) has appeared on the market and can replace the old lacquers which contain solvents like alcohols and aromatic solvents, as well as formaldehyde.

If hydrochloric acid is used for hardening that type of parquet lacquer, there is a risk of the formation of bis-chloromethyl ether. In general, however, organic acids are used, such as para-toluene sulphonic acid, eliminating risk. Parquet floors were the last group of frequently-occurring large surfaces that used to require coatings based on organic solvents.

Paints contain many other things besides solvents. They contain binder, pigment, filler, and various additives. Here, too, the growing knowledge about health hazards has led to numerous improvements. Many components have now been eliminated from paints: fungicides based on mercury, pentachloro-phenol or plasticizers like PCB (polychloro-biphenyls); pigments based on cadmium; and fillers like asbestos.

I would like to mention the case of lead chromate pigments, an improvement that has only arrived half way. They are used in many bright shades of yellow, orange, and red paint. Although a number of substitutes does exist, we would be very glad if we *could* avoid the lead pigments. The trouble is that the substitutes are not yet fully accepted. They have poor hiding power and so have not yet won the approval of painters. We must wait until health demands become so insistent that we are prepared to apply a second coat of paint in order to ensure adequate hiding ability, or, better yet, modern research programs will yield harmless pigments of the proper colors and with great hiding power.

Good products that are not hazardous—manufacturers and users are all interested! They want to see them developed and put into use. But remember that we shall never get products that are completely harmless. So it is necessary to use wise judgment when applying paints, and to take special health precautions in some cases. Epoxy paints are an example. They cannot be replaced by other products that are completely harmless, so an extensive training program has been worked out for people who work with epoxy products. It enables them to handle them with minimum risk and all those who use epoxy products in their profession must complete this course.

There is one important sector still needing development in the future. It involves fast, reliable methods of evaluating new chemicals. The aim is to avoid hazardous substances before they come into use, or else to ensure suitable health precautions.

A great deal of developmental work has been done on paints in the last ten to fifteen years. Much of this work has involved improving the health aspects and replacing existing paints with substitutes that are less hazardous. Every change has demanded extensive research and development. The research efforts have led to progress and the manufacturers' products now have better properties, too, which inevitably means that the new, less hazardous components and products are often more expensive. However, the overall effect of such higher prices is relatively unimportant.

SECTION II

CHAPTER 5

EXPERIMENTAL APPROACH TO THE ASSESSMENT OF THE CARCINOGENIC RISK OF INDUSTRIAL INORGANIC PIGMENTS

Cesare Maltoni, Leonildo Morisi and Pasquale Chieco
Institute of Oncology
Bologna, Italy

INTRODUCTION

The quantity and the number of natural or industrially produced inorganic compounds are progressively increasing. Conversely, there has not been a parallel intensification of scientific efforts to study their potential adverse effects on health and, particularly, to assess which agents may represent an oncogenic risk. This is especially true for inorganic pigments.

The available data on the carcinogenic effects of inorganic pigments consist of a few past epidemiological investigations made on heavily exposed workers, and some early experimental research, now mainly of historical interest. Therefore, there is presently a need for systematic epidemiological and experimental studies in this field.

The following population groups may be exposed to inorganic compounds and, among them, to inorganic pigments:
—workers engaged in their production;
—workers manufacturing products containing them;
—workers using products in which they are contained;
—and, to a lesser extent, residents near factories in which they are produced or used;
—and people in general undergoing contact with them.

Several years ago we started a large integrated project of long-term bioassays on different inorganic compounds, the majority of which were pigments.

The aims of this project were:
(1)   to screen a series of compounds, most of which have never been studied, to identify the carcinogenic ones;
(2)   to operate in a highly standardized condition so as to obtain information on the relative oncogenic potency of the ones which would turn out to be carcinogens.

As a test for our long—term bioassays we used injection of the compounds in subcutaneous tissues. This test is appropriate for compounds not

TABLE 1

Background incidence of SUBCUTANEOUS SARCOMAS among Sprague-Dawley rats, at the Institute of Oncology of Bologna (Historical Controls)

Group	Treatment	In ANIMALS				In ANIMALS WITH SARCOMAS Histotype			
		Sex	No. at start	No.	%	Fibro-sarcomas	Rhabdomyo-sarcomas	Lipo-sarcomas	Others
I	None	M	1,434	5	0.3	5	0	0	0
		F	1,457	4	0.3	0	2	0	2
		M and F	2,891	9	0.3	5	2	0	2
II	Oncogenic and non-oncogenic agents administered by general routes (inhalation, ingestion, intraperitoneal injection)	M	4,015	18	0.4	13	3	2	0
		F	4,101	14	0.3	8	4	0	2
		M and F	8,116	32	0.4	21	7	2	2
III	Subcutaneous injection of saline	M	145	0	-	0	0	0	0
		F	115	0	-	0	0	0	0
		M and F	260	0	-	0	0	0	0
IV	Subcutaneous injection of olive oil	M	75	1	1.3	1	0	0	0
		F	80	2	2.5	1	1	0	0
		M and F	155	3	1.9	2	1	0	0
V	Subcutaneous injection of indirect carcinogens in olive oil	M	75	1	1.3	0	1	0	0
		F	80	0	-	0	0	0	0
		M and F	155	1	0.6	0	1	0	0
Total			11,577	45	0.4	28	11	2	4

easily diffusable in tissues, and for potential carcinogens, not needing biotransformation to produce the carcinogenic effects locally (direct carcinogens). This is the case of many inorganic chemical agents.

In the past, it was claimed that sarcomas of subcutaneous tissues might be produced in rats by compounds generally considered innocuous (hypertonic solutions, sugar, olive oil, etc.) and that, therefore, the test would not be suitable for carcinogenic bioassays.

After 25 years of experience we can say that this claim is not true. The subcutaneous tumours observed in the past, in rats injected with innocuous compounds, were probably spontaneously occurring, since the onset of subcutaneous sarcomas is not infrequent in rats of some strains, particularly in old age. Moreover, the spontaneously arising sarcomas are usually very well differentiated, highly monomorphic fibrosarcomas, originating from mammary gland stroma, while the sarcomas originating at the site of injection of a true carcinogen are usually more or less differentiated rhabdomyosarcomas rising from the *panniculus carnosus*, and less frequently, polymorphic fibrosarcomas.

MATERIALS AND METHODS

In all the experiments 30 mg of the compounds dissolved and/or suspended in saline, were injected *una tantum* in the subcutaneous tissue of the middle right flank of male and female Sprague-Dawley rats, 13 weeks old at the start of the experiments.

The spontaneous incidence of subcutaneous sarcomas and of the different sarcoma histotypes, in the historical controls of our breed of Sprague-Dawley rats, are reported in Table 1.

Rats receiving subcutaneous injections of saline were used as controls. All the animals were kept under observation until spontaneous death.

During the experiments, the animals were examined and weighed every two weeks. All macroscopic lesions observed at the control were recorded. A complete autopsy was made on each animal. Histological examinations were performed on the tumours and tissues at the site of injection, on the major organs (brain, thymus and mediastinal nodes, lung, liver, spleen, pancreas, kidneys and adrenals, stomach, uterus, gonads, and subcutaneous and mesentheric nodes) and any other organ with pathological lesions.

The experimental design and results are reported in Tables 2–7.

RESULTS

The observed tumours at the site of injection, are mainly rhabdomyosarcomas (more or less differentiated) (Fig. 1-6), fibrosarcomas (Fig. 7, 8) and mixed sarcomas (rhabdo-fibrosarcomas).

From the present data it can be seen that:

TABLE 2

Carcinogenicity bioassays of CHROMIUM YELLOW (LEAD CHROMATE) and CHROMIUM ORANGE (BASIC LEAD CHROMATE), by a single subcutaneous injection (30 mg in 1 ml of saline) in Sprague-Dawley rats. (Control rats treated with a single injection of 1 ml of saline).

Experiment No.	Compound	No. of Animals at start			Duration of the experiment (in weeks)	Animals with Sarcomas at the site of injection				Histological type of local sarcomas
		M	F	Total		M	F	Total		
						No.	No.	No.	%	
1	Chromium Yellow (Lead Chromate)	20	20	40	150	10	16	26	65	Rhabdomyosarcomas and fibrosarcomas
2	Chromium Orange (Basic Lead Chromate)	20	20	40	132	14	13	27	67	Rhabdomyosarcomas and fibrosarcomas
	Control	45	15	60	124	0	0	0	-	-

TABLE 3

Carcinogenicity bioassays of ZINC YELLOW (BASIC ZINC CHROMATE), by a single subcutaneous injection (30 mg in 1 ml of saline) in Sprague–Dawley rats.
(Control rats treated with a single injection of 1 ml of saline).

Experiment No.	Compound	No. of Animals at start			Duration of the experiment (in weeks)	Animals with Sarcomas at the site of injection				Histological type of local sarcomas
		M	F	Total		M No.	F No.	Total No.	Total %	
1	Zinc Yellow (Basic Zinc Chromate) (1)	20	20	40	110	3	3	6	15	Rhabdomyosarcomas and fibrosarcomas
2	Zinc Yellow (Basic Zinc Chromate) (2)	20	20	40	137	9	8	17	42	Rhabdomyosarcomas and fibrosarcomas
	Control	20	20	40	136	0	0	0	-	-

(1) C_2O_3: 20%
(2) C_2O_3: 40%

TABLE 4

Carcinogenicity bioassays of MOLYBDENUM ORANGE (LEAD CHROMATE, SULPHATE and MOLYBDATE), by a single subcutaneous injection (30 mg in 1 ml of saline) in Sprague-Dawley rats.
(Control rats treated by a single injection of 1 ml of saline).

Experiment No.	Compound	No. of Animals at start			Duration of the experiment (in weeks)	Animals with Sarcomas at the site of injection					Histological type of local sarcomas
		M	F	Total		M	F	Total			
						No.	No.	No.	%		
1	Molybdenum Orange (Lead Chromate, Sulphate and Molybdate)	20	20	40	117	19	17	36	90		Rhabdomyosarcomas and fibrosarcomas
	Control	45	15	60	124	0	0	0	-		-

TABLE 5

Carcinogenicity bioassays of CADMIUM YELLOW (CADMIUM SULPHIDE), by a single subcutaneous injection (30 mg in 1 ml of saline) in Sprague-Dawley rats. (Control rats treated with a single injection of 1 ml of saline).

Experiment No.	Compound	No. of Animals at start			Duration of the experiment (in weeks)	Animals with Sarcomas at the site of injection					Histological type of local sarcomas
		M	F	Total		M	F	Total			
						No.	No.	No.	%		
1	Cadmium Yellow (Cadmium Sulphide)	20	20	40	143	9	7	16	40	Rhabdomyosarcomas and fibrosarcomas
	Control	45	15	60	124	0	0	0	-	-

TABLE 6

Carcinogenicity bioassays of IRON YELLOW and IRON RED (IRON OXIDE), by a single subcutaneous injection (30 mg in 1 ml of saline) in Sprague–Dawley rats. (Control rats treated with a single injection of 1 ml of saline).

Experiment No.	Compound	No. of Animals at start			Duration of the experiment (in weeks)	Animals with Sarcomas at the site of injection					Histological type of local sarcomas
		M	F	Total		M	F	Total			
						No.	No.	No.	%		
1	Iron Yellow (Iron Oxide)	20	20	40	141	0	0	0	-	-
2	Iron Red (Iron Oxide)	20	20	40	134	1	0	1	2	Rhabdomyosarcoma and fibrosarcoma
	Control	45	15	60	124	0	0	0	-	-

TABLE 7

Carcinogenicity bioassays of TITANIUM OXIDE, by a single subcutaneous injection
(30 mg in 1 ml of saline) in Sprague-Dawley rats.
(Control rats treated with a single injection of 1 ml of saline).

Experiment No.	Compound	No. of Animals at start			Duration of the experiment (in weeks)	Animals with Sarcomas at the site of injection				Histological type of local sarcomas
		M	F	Total		M	F	Total		
						No.	No.	No.	%	
1	Titanium Oxide (1)	20	20	40	126	0	0	0	-	-
2	Titanium Oxide (2)	20	20	40	146	0	0	0	-	-
3	Titanium Oxide (3)	20	20	40	133	0	0	0	-	-
	Control	20	20	40	136	0	0	0	-	-

(1) K_2O : 0.3% ; P_2O_5 : 0.3% ; Sb_2O_3 : 0.2%.
(2) Al_2O_3 : 3.3% ; K_2O : 0.2% ; P_2O_5 : 0.1% ; $Si O_2$: 0.7%.
(3) Al_2O_3 : 3.8% ; K_2O : 0.2% ; P_2O_5 : 0.2% ; Sb_2O_3 : 0.3% ; $Si O_2$: 10.5%.

TABLE 8

Frequency of ATYPICAL ADENOMATOUS HYPERPLASIA and of SQUAMOUS DYSPLASIA among WORKERS EXPOSED TO CHROMIUM COMPOUNDS

Occupation	Years of Exposure	No. of Workers	Atypical Adenomatous Hyperplasia		Squamous Dysplasia	
			No.	%	No.	%
Production of chromates and bichromates	<5	25	0	-	4	16
	6-10	14	1	7	3	21
	11-15	13	3	23	4	30
	>15	17	1	6	4	23
	Total	69	5	7	15	22
Production of chromium pigments	<5	14	0	-	0	-
	6-10	17	0	-	2	12
	11-15	8	0	-	1	12
	>15	8	0	-	2	25
	Total	47	0	-	5	10

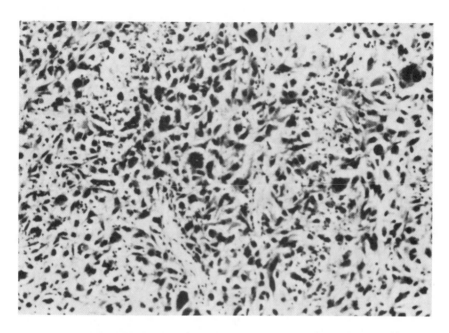

FIGURE 1 Well differentiated rhabdomyosarcoma. H.E. x 128.

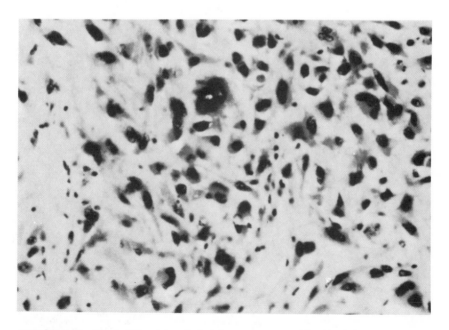

FIGURE 2 Well differentiated rhabdomyosarcoma (detail of fig. 1). H.E. x 320.

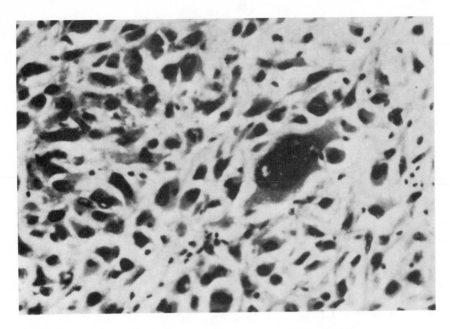

FIGURE 3   Well differentiated rhabdomyosarcoma. H.E. x 320.

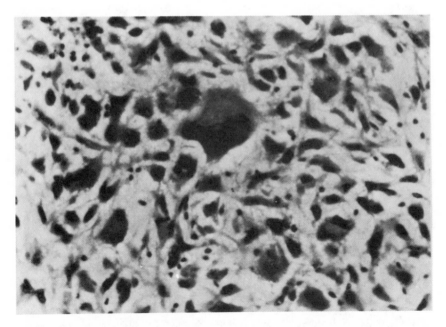

FIGURE 4   Well differentiated rhabdomyosarcoma. H.E. x 320.

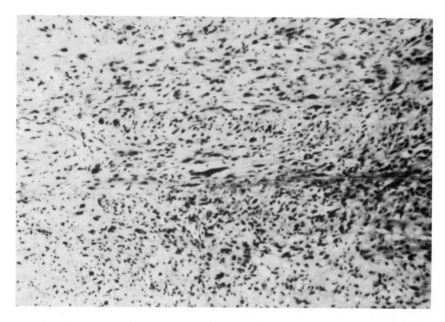

FIGURE 5 Poorly differentiated rhabdomyosarcoma. H.E. x 128.

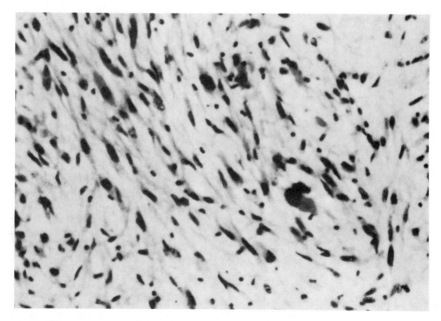

FIGURE 6 Poorly differentiated rhabdomyosarcoma. H.E. x 320.

FIGURE 7   Fibrosarcoma. H.E. x 128.

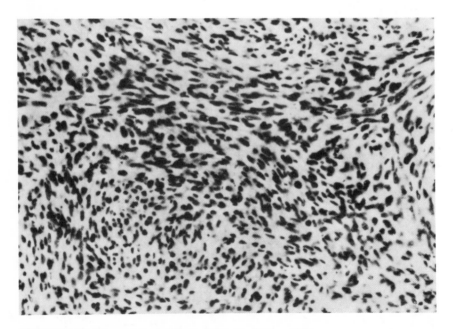

FIGURE 8   Fibrosarcoma (detail of fig. 7). H.E. x 320.

A) Chromium Yellow, Chromium Orange, Zinc Yellow of two types, Molybdenum Orange and Cadmium Yellow show a definite carcinogenic effect.
B) Iron Oxide shows a border-line effect (1 sarcoma observed in 80 animals).
C) Titanium Oxide does not show any oncogenic effects.
D) The greatest carcinogenic effect was observed for Chromium Yellow and Chromium Orange, and for Molybdenum Orange. This last compound appears to be, under our experimental conditions, one of the most potent carcinogens ever tested by subcutaneous route.

CONCLUSIONS AND DISCUSSION

Inorganic pigments currently widely-produced and used, may turn out to represent a real carcinogenic risk for humans. Therefore, this field requires more scientific action as well as greater control by public health authorities.

As urgent steps we suggest:
A) conducting systematic experimental research to identify carcinogenic compounds and to assess their level of risk;
B) implementing epidemiological investigations over the exposed population groups;
C) performing medical surveillance on risk groups. In our opinion these controls are particularly necessary for painters, who may be heavily exposed to the pigments we have shown to be highly carcinogenic. For such controls, since the respiratory tract is the most highly exposed organ, we recommend sputum cytology as a tool not only as a control of health, but for monitoring the risk on a social scale.

In our Institute we are currently using sputum cytology, with good results (Table 8).

The validity of the experimental bioassays in pointing out the oncogenic risk represented by pigments is clearly demonstrated by recent facts.

The results of the experimental bioassay showing the carcinogenic potential of Chromium Yellow and Orange, were published in 1973. More than two years later, i.e. in October 1975, the Dry Color Manufacturers Association released the data from an epidemiological investigation, which has been started after publication of our experimental evidence of 580 workers exposed to lead chromate. These data showed that lung cancer was the cause of death for nearly 29% of the deceased workers, and accounted for 85% of all cancer deaths.

SUMMARY

Long-term carcinogenicity bioassays on several inorganic pigments, namely Chromium Yellow, Chromium Orange, Zinc Yellow, Molybdenum Orange, Cadmium Yellow, Iron Yellow, Iron Red and Titanium Oxide, were undertaken. The compounds were injected, *una tantum*, in the subcutaneous tissues of Sprague-Dawley rats.

Molybdenum Orange, Chromium Orange and Yellow, Zinc Yellow and

Cadmium Yellow, were found to be markedly oncogenic (in decreasing order of potency). Border-line effects were observed for Iron Red. No effect was detected for Iron Yellow and Titanium Oxide.

The need for the extension of experimental bioassays to other inorganic pigments, of epidemiological investigations and medical surveillance, and of preventive measures, is emphasized.

CHAPTER 6

CHEMICAL HAZARDS IN PAINTING IN THE CONSTRUCTION INDUSTRY

Riitta Riala, Ind. Hyg.
Uusimaa Regional Institute of Occupational Health
Arinatie 3 A, SF-00370 Helsinki 37

INTRODUCTION

It is estimated that 60% of all paints used in the construction industry in Finland are water-based. In new buildings, the use of solvent-containing paints is relatively rare, but in rebuilding and in repair work, as well as in construction of factories, they are still used to a large extent.

A Survey on Construction Work

During the years 1974–77 the Institute of Occupational Health performed a study on work conditions in the construction industry. The study consisted of ergonomic and industrial hygiene sections. Industrial hygiene measurements were performed on a new building site also during painting and varnishing work.

The solvent exposure of painters was measured by sampling the vapors in the painter's breathing zone in charcoal tubes and analyzing the samples in the laboratory by gas chromatography. Formaldehyde vapors during carbamide varnishing were absorbed in water and analyzed by chromotopic acid method.

Solvent-containing alkyd paints were used in the surveyed construction site only for small surfaces, e.g. paintings of radiators, stair flight railings, etc. Alkyd varnishes and wood preservaties were also used in small amounts, e.g. in the painting of window frames, sauna platforms, etc. The Stoddard solvent concentrations measured during these paintings were relatively low (Fig. 1).

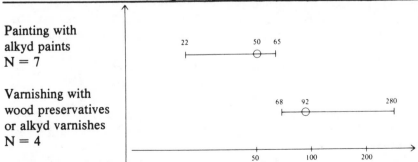

FIGURE 1 Solvent concentrations during painting and alkyd varnishing. N = number of measurements. O = median value.

The parquet floors were varnished with two types of varnishes: The undercoat, nitro-cellulose lacquer had as main solvents acetone, ethanol and butyl acetate. As a finish ureformaldehyde varnish, the solvents of which are ethanol and isobutanol was used. The solvent concentrations during varnishing work have been compared to Finnish and Swedish TLV's (Fig. 2). The solvent concentrations can often exceed the TLV's during varnishing.

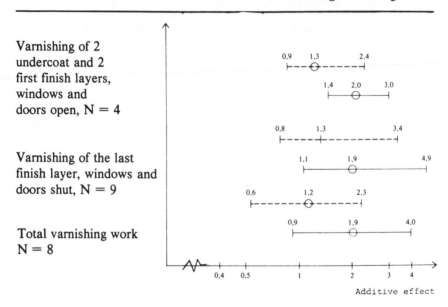

Varnishing of 2
undercoat and 2
first finish layers,
windows and
doors open, N = 4

Varnishing of the last
finish layer, windows and
doors shut, N = 9

Total varnishing work
N = 8

Additive effect

FIGURE 2 Exposure to solvents during floor varnishing. N = number of measurements. O= median value.
_ _ _ _ _ additive effect according to Finnish TLV's (1972)
_____ additive effect according to Swedish TLV's (1978)

During the use of urea formaldehyde varnishes, formaldehyde can be released. The average formaldehyde concentration of 10 air samples taken during varnishing work was 2,8 cm³/m³. The average of 26 Dräger test tube measurements during the application of the last varnish layer was 4,5 cm³/m³.

A Study on Health Hazards of Concrete Reinforcement Workers and Painters

In a study that was stated in 1977, the health hazards of the work of concrete reinforcement workers and painters are surveyed. The two occupational groups are referents for each other.

The study includes orthopedic, psychological and neurophysiological, ergonomic and industrial hygiene sections. In the industrial hygiene study, the exposure of repair painters to paint solvents in different work conditions is investigated. By questionnaires the earlier exposure of painters is surveyed. By industrial hygiene measurements the solvent concentration during different painting situations is investigated. The study was finished and reported in 1981.

REFERENCES

1. Ahonen, K., Engström, B., Järvenpää, E., Lehtinen, P.U., Saari, J. & Wickström, G.: Survey of construction work: Part I. General review. Työolosuhteet 6. Institute of Occupational Health, Helsinki 1977. 197 p. (Finnish)

2. Ahonen, K., Engström, B., & Lehtinen, P.U.: Survey of construction work: Part II. Occupational hygiene. Työolosuhteet 7. Institute of Occupational Health, Helsinki 1977. 73 p. + appendices (Finnish with Swedish Summary)

CHAPTER 7

EARLY RESULTS OF THE EXPERIMENTAL ASSESSMENTS OF THE CARCINOGENIC EFFECTS OF ONE EPOXY SOLVENT: STYRENE OXIDE

Cesare Maltoni
Institute of Oncology
Bologna, Italy

INTRODUCTION

Styrene oxide, or styrene epoxide, (epoxyethyl) benzene (molecular weight 122) is a colorless-to-pale-straw-coloured liquid, with a boiling point of 194.1°C. It is produced commercially either by the chlorohydrin route or by epoxidation of styrene with peroxiacetic acid. The monomer was first produced on an industrial scale in Japan in 1964 and commercial production in the USA was first reported in 1974. It is also industrially produced now in Europe. The world-wide production of styrene oxide is estimated at about 3,000 tons per year.

It is used as a reactive diluent in epoxy resins to reduce the viscosity of mixed systems prior to curing, as an intermediate in the preparation of agricultural and biological chemicals, cosmetics, and surface coating, in the treatment of textile and fibers, and as a raw material for the production of phenylstearyl alcohol used in perfume. In the USA, the FDA has ruled that styrene oxide may be used as a catalyst and cross-linking agent for epoxy resins in coating for containers, having a capacity of 1,000 gallons (3,785 l) or more, when such containers are intended for repeated use in contact with beverages containing up to 8% alcohol by volume (USA Food and Drug Administration, 1977).

The following population groups may be exposed to styrene oxide:
—workers engaged in the production of styrene oxide and its derivatives,
—workers manufacturing and using products containing styrene oxide,
—and, to a lesser extent, residents near factories producing or using styrene oxide,
—and people, in general, coming into contact with products containing styrene oxide.

There were no studies prior to the present one on the long-term general effects of styrene oxide in experimental animals, and there are no available epidemiological studies on exposed populations. The only long-term tests are

Experiment BT105:   Exposure by ingestion (stomach tube) to Styrene Oxide, in olive oil, at 250 and 50 mg/Kg b.w., once daily, 4-5 days weekly, for 52 weeks. Results after 156 weeks (end of experiment).

TABLE 1

DISTRIBUTION OF THE FORESTOMACH EPITHELIAL TUMOURS (BENIGN AND MALIGNANT)

GROUP NO.	CONCEN-TRATION	ANIMALS (Sprague-Dawley rats, 13 weeks old at start)			ANIMALS WITH FORESTOMACH EPITHELIAL TUMOURS (b)			Histotype									
								Papillomas and acanthomas			Squamocellular carcinomas						
											Total			Extension			
														In situ		Invasive	
		Sex	No. at start	Correct-ed number (a)	No.	% (c)	Average latency time (weeks) (d)	No.	% (c)	Average latency time (weeks) (d)	No.	% (c)	Average latency time (weeks) (d)	No.	% (c)	No.	%
I	250 mg/Kg	M	40	39	19	48.7	107.7	7	17.9	109.1	16	41.0	110.9	16	41.0	8	20.5
		F	40	38	21	55.3	106.8	5	13.1	107.2	20	52.6	105.3	16	42.1	10	26.3
		Total	80	77	40	51.9	107.2	12	15.6	108.3	36	46.7	107.8	32	41.5	18	23.4
II	50 mg/Kg	M	40	39	10	25.6	100.3	3	7.7	108.0	9	23.1	104.9	6	15.4	5	12.8
		F	40	37	7	18.9	121.1	2	5.4	130.0	7	18.9	121.1	7	18.9	1	2.5
		Total	80	76	17	22.4	108.9	5	6.6	116.8	16	21.0	112.0	13	17.1	6	7.9
III	Olive oil (Control)	M	40	39	0	-	-	0	-	-	0	-	-	0	-	0	-
		F	40	40	0	-	-	0	-	-	0	-	-	0	-	0	-
		Total	80	79	0	-	-	0	-	-	0	-	-	0	-	0	-
Total			240	232													

(a) Alive animals after 17 weeks, when the first forestomach epithelial tumour was observed.
(b) One or more tumours may be present in the same animal.

(c) The percentages are referred to corrected numbers.
(d) Average time from the start of the experiment.

two carcinogenic studies on mice by skin application, in which no increase in the incidence of cutaneous tumours was observed (Van Duuren et al., 1963; Weil et al., 1963).

In 1976 we started a long-term carcinogenicity bioassay on styrene oxide. There were three main reasons for this experiment:
1) the volume of production and the variety of uses of this compound called for information on possible long-term risks with particular regard to carcino-genicity,
2) the fact that the epoxy-derivatives belong to a suspicious category, and
3) the hypothesis that styrene oxide is the possible active metabolite of styrene, which, in turn, is one of the most important monomers used in the plastics industry.
Early results of this experiment have been previously published (Maltoni et al., 1979).

MATERIAL AND METHODS

The experiment was designed to study the long-term effects of styrene oxide, administered by ingestion. The compound was kindly supplied by a producer. As a vehicle, pure olive oil was used.

The experiment utilized Sprague-Dawley rats, 13 weeks old at the beginning of the experiment, of the same breed currently used for bioassays in our Institute. The animals were weaned and classified by sex when 4–5 weeks old, at which time they were numbered by ear punch and divided into groups by litter distribution. After weaning, the animals were fed, *ad libitum*, an adequate commercial diet. The animals were kept in groups of five in (R)macrolon* cages, with tops of stainless steel wire. A shallow layer of white wood shavings served as bedding. The animals were kept in a temperature-controlled laboratory at 22°±3°C.

The compound was administered in an oil solution by stomach tube once daily, 4–5 days weekly, for 52 weeks. Two doses were tested: 250 mg/Kg b.w. and 50 mg/Kg b.w. The animals of the control group were treated with olive oil alone. The plan of the experiment is given in Table 1.

After the end of the treatment period, the animals were allowed to live until spontaneous death.

During the experiment the animals were examined every two weeks; they were weighed every two weeks during the period of treatment, and then every eight weeks.

All detectable gross pathological changes were recorded during the examination. All the animals were kept under observation until spontaneous death. The animals, when moribund, were isolated, in order to avoid cannibalism.

A complete autopsy was made on each animal. Histological examinations were performed on Zymbal glands, interscapular brown fat, salivary glands, tongue, thymus and mediastinal nodes, lung, liver, kidneys, adrenals, spleen,

*polycarbonate

TABLE 2

Experiment BT105: Exposure by ingestion (stomach tube) to Styrene Oxide, in olive oil, at 250 and 50 mg/Kg b.w., once daily, 4–5 days weekly, for 52 weeks. Results after 156 weeks (end of experiment).

DISTRIBUTION OF THE FORESTOMACH PRECURSOR LESIONS

GROUP NO.	CONCEN-TRATION	ANIMALS (Sprague-Dawley rats, 13 weeks old at start) Sex	No. at start	ANIMALS WITH PRECURSOR LESIONS No.	% (a)	Average latency time (weeks) (b)	Simple hyperplasia No.	Simple hyperplasia % (a)	Acantho-matosis No.	Acantho-matosis % (a)	Squamous dysplasia No.	Squamous dysplasia % (a)
I	250 mg/Kg	M	40	14	35.0	96.3	0	-	3	7.5	11	27.5
		F	40	12	30.0	75.2	2	5.0	4	10.0	6	15.0
		Total	80	26	32.5	86.6	2	2.5	7	8.7	17	21.2
II	50 mg/Kg	M	40	5	12.5	98.0	0	-	1	2.5	4	10.0
		F	40	7	17.5	104.1	0	-	3	7.5	5	12.5
		Total	80	12	15.0	101.6	0	-	4	5.0	9	11.2
III	Olive oil (Control)	M	40	1	2.5	69.0	0	-	1	2.5	0	-
		F	40	2	5.0	105.5	0	-	1	2.5	1	2.5
		Total	80	3	3.7	93.3	0	-	2	2.5	1	1.2
Total			240									

(a) The percentages are referred to the number at start.
(b) Average time from the start of the experiment.

stomach, different segments of the intestine, bladder, uterus, gonads, brain and any other organ with pathological lesions.

RESULTS

The experiment lasted 156 weeks. The data reported here should be considered preliminary, since they are solely concerned with the oncological lesions of the stomach which, under our experimental conditions, appears to be the target organ. The final report of all the results obtained will be presented *in extenso* in the near future.

Acanthomas, papillomas and *in situ* and invasive squamous carcinomas of the forestomach were observed at the two studied dose levels, with a clear-cut dose-response relationship. Invasive carcinomas often metastatize to the liver. More than one of these tumours may be observed in the same animal, in different parts of the organ. In the forestomach of the treated animals, with or without tumours, one or more precursor lesions—i.e. simple hyperplasia, acanthomatosis and squamous dysplasia—were frequently found. Examples of the various observed preneoplastic and neoplastic lesions, are presented in Figures 1–16.

The incidence of the different types of gastric neoplasias and precursor lesions and their latency period are reported in Tables 1 and 2, respectively.

It should be pointed out that epithelial tumours of the forestomach are rare in the colony of rats used in our laboratory. Among the historical control rats of our laboratory, kept under the same conditions, and handled and examined in the same standard way, we found the following incidence: 27 benign epithelial tumours (papillomas and acanthomas) among 2,376 untreated rats; 9 benign epithelial tumours and 1 carcinoma among 644 rats, treated by ingestion (stomach tube) with a daily dose of 0.6–1.2 ml of olive oil, given 4–5 days weekly, for 52–59 weeks.

CONCLUSIONS AND DISCUSSION

On the basis of the presented results, styrene oxide appears to be a very potent, direct carcinogen: in fact, it strongly affects the organ most directly exposed.

It may be reasonably presumed that, with other routes of exposure, other organs and tissues more intensely exposed may also be affected.

In our opinion, further experiments should be undertaken by inhalation, using a wider range of doses, in order to better assess the levels of risk.

Meanwhile, epidemiological investigations and medical surveillance should be performed in occupational exposed groups, and preventive measures promptly considered.

Moreover, in the light of these data, although the results of long-term bioassays of styrene in rats performed by us at the Institute of Oncology of Bologna failed to show clear carcinogenic effects under the considered

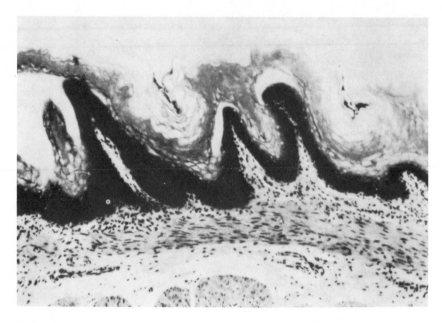

FIGURE 1 Forestomach acanthomatosis. H.E. x 100.

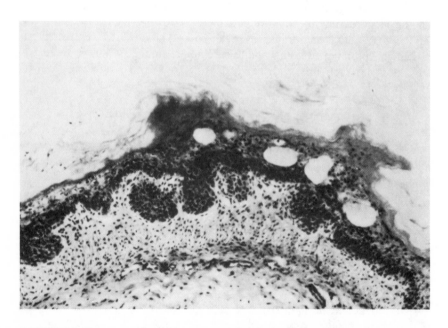

FIGURE 2 Forestomach early squamous dysplasia. H.E. x 128.

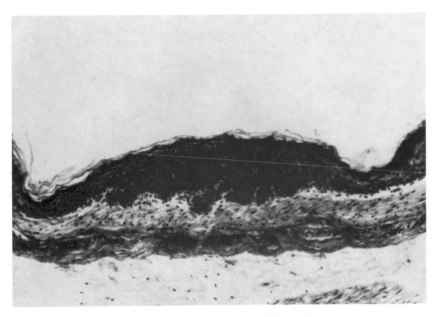

FIGURE 3 Forestomach squamous dysplasia. H.E. x 128.

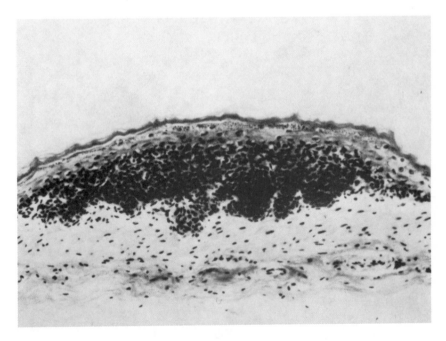

FIGURE 4 Forestomach squamous dysplasia (detail of fig. 3). H.E. x 200.

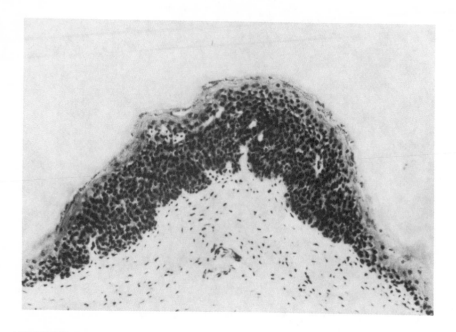

FIGURE 5 Forestomach grave squamous dysplasia. H.E. x 200.

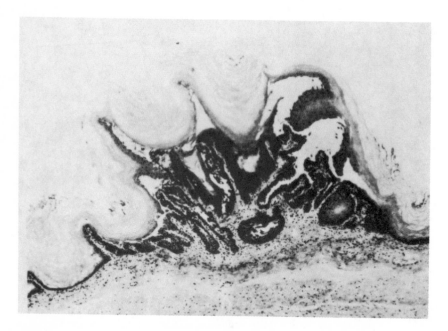

FIGURE 6 Forestomach acanthoma. H.E. x 80.

FIGURE 7 Forestomach papilloma. H.E. x 32.

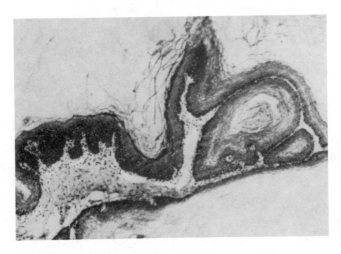

FIGURE 8 Forestomach papilloma, with an area of marked dysplasia-in situ squamous carcinoma (on left side). H.E. x 80.

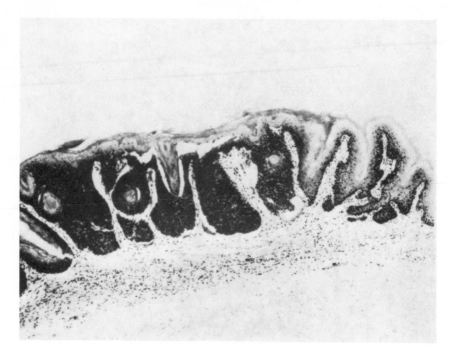

FIGURE 9 Forestomach in situ squamous carcinoma. H.E. x 80.

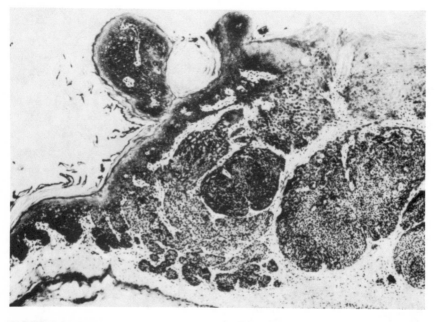

FIGURE 10 Forestomach early invasive (micro-invasive) squamous carcinoma. H.E. x 80.

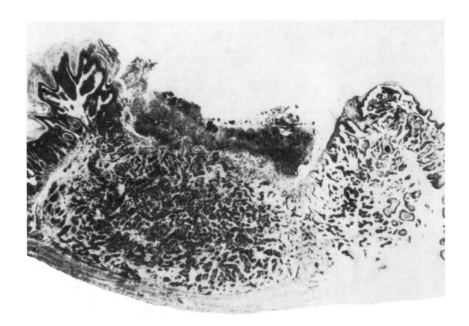

FIGURE 11 Forestomach invasive squamous carcinoma. H.E. x 20.

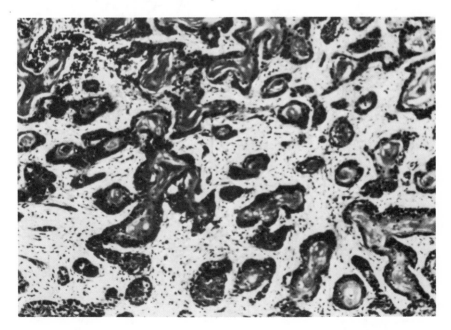

FIGURE 12 Forestomach invasive squamous carcinoma (detail of figure 11). H.E. x 128.

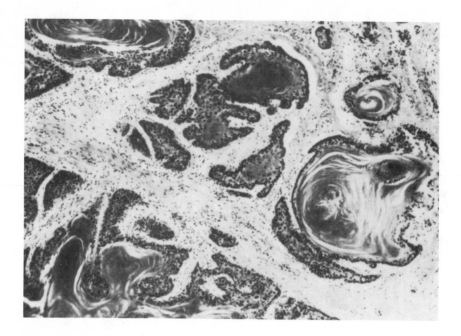

FIGURE 13 Forestomach invasive squamous carcinoma. H.E. x 80.

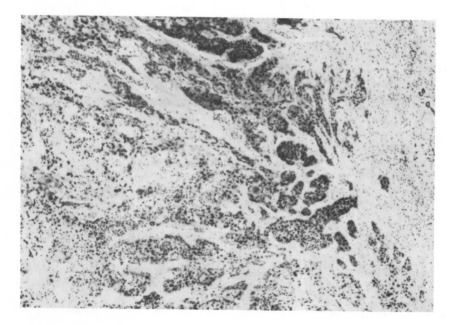

FIGURE 14 Liver metastasis of the forestomach carcinoma shown in fig. 13 (biliary ducts are visible on the right side). H.E. x 80.

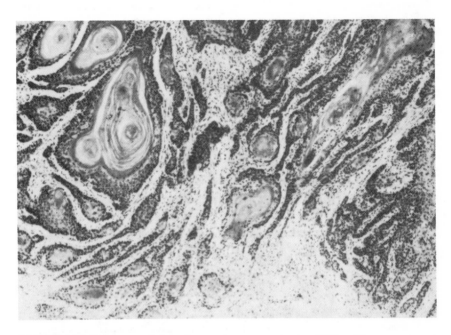

FIGURE 15 Forestomach invasive squamous carcinoma. H.E. x 80.

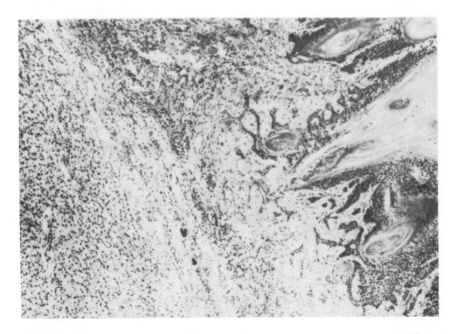

FIGURE 16 Liver metastasis of the forestomach carcinoma shown in fig. 15 (liver parenchyma is visible on the left side). H.E. x 80.

experimental conditions (Maltoni et al., in press), the entire matter of the safety of styrene must be more carefully reconsidered.

SUMMARY

Long-term carcinogenicity bioassays on styrene oxide were undertaken. The monomer was tested by ingestion by stomach tube, at 2 dose levels, 250 and 50 mg/Kg b.w., on Sprague-Dawley rats. The present report deals with the effects of the compound on the stomach (forestomach), which seems to be the target organ, under our experimental conditions. A high incidence of papillomas and acanthomas and of carcinomas of the forestomach were observed in the treated animals at both doses, with a clear-cut dose-response relationship. To our knowledge, this study is the first on chronic toxicity of styrene oxide and represents the first indication of its carcinogenicity.

REFERENCES

Maltoni C., Failla G., and Kassapidi G.: First experimental demonstration of the carcinogenic effects of styrene oxide. Long term bioassays on Sprague-Dawley rats by oral administration. *La Medicina del Lavoro* 70: 358, 1979.

Van Duuren B.L., Nelson N., Orris L., Palmes E.D. and Schmitt F.L.: Carcinogenicity of epoxides, lactones, and peroxy compounds. *J. Nat. Cancer Inst.* 31: 41, 1963.

Weil C.S., Condra N., Haun C. and Striegel J.A.: Experimental carcinogenicity and acute toxicity of representative epoxides. *Am. Ind. Hyg. Assoc. J.* 24: 305, 1963.

CHAPTER 8

HEALTH HAZARDS AMONG PAINTERS

Knut Ringen, Dr. P.H.

Workers' Institute for Safety and Health
Washington, D.C.

SUMMARY

The health of painters has been of great concern in recent years, and OSHA now considers painting to be an occupational high health risk. As a result, a large body of data on health related aspects of painting is appearing. This report summarizes the main conclusions that can be drawn from a review of the available information, and proposes a number of priorities that could be implemented through relatively short-term policy initiatives.

This report has two major themes. First, the painters' work environments today are largely uncontrolled, particularly in the construction industry. Areas of consideration to improve controllability include: a) increase compliance with existing regulations, including the use of on-site government inspectors at worksites; b) provide government subsidies to small businesses to assist in developing improved health protection; and c) encourage certification of painters in especially hazardous occupations.

Second, while it is recognized that some workplaces can be controlled through engineering of self-contained painting areas, the majority of workplaces for painters cannot be controlled adequately to ensure acceptable protection to the workers. To protect these painters, exposure to a small number of substances should be reduced immediately, and the feasibility of removing the substances from the market should be studied.

Critical substances at this time include:

Pigments and fillers: lead, chromium, mercury, asbestos, talc

Solvents: commercial grade aromatic hydrocarbons (toluene, xylene, styrene/ethyl benzyne), methyl n-butyl ketone, n-hexane, isobutanol, and acetone

Resins: epoxy.

Due to their possible health hazards and extensive use, the pigment titanium dioxide and the resins polyurethanes and acrylics should be studied more fully to determine long-term health effects as sources of exposure to painters.

A. INTRODUCTION

Concern about occupational health, while having a long history, has intensified greatly in the last two decades. Remarkable in this development is the increased concern about chronic disease causation. While the reasons for this development can be found in a very complex social dialectic, three critical factors prevail: (1) increased evidence of chronic diseases (and concomitant decline of acute infectious diseases); (2) improved scientific methods (analytical chemistry, clinical investigation, epidemiology, engineering and industrial hygiene); and (3) growing political constituencies in public health.

Until recently, occupational health centered around protecting workers from acute health effects. Today, chronic diseases are of equal concern. Among chronic health concerns, cancer has been emphasized while other chronic diseases, including neurotoxicity, heart diseases and stress, have received less attention. However, in spite of the efforts made even in the case of cancers, the etiologic mechanisms are not understood. Generally, it appears that these diseases are dose-related, but that levels of safe exposure (e.g., thresholds) are not known.[1] Also, a multiplicity of exposures to many factors, some inside the workplace and some found in the community environment and personal lifestyles, appear to combine in the etiology of these diseases and increase the risk of disease.[2,3] Invariably, long latency periods are involved; the generation of a detectable disease effect may take from 5-40 years.[4-5] Because these diseases "creep up" over a long period of time, they are difficult to detect early, and the traditional concepts of "incubation period" and "incidence'" of infectious disease epidemiology may not be very meaningful in dealing with chronic diseases.

Protection of health from chronic diseases associated with environmental factors is thus fraught with complexity and scientific uncertainty. This has been opportune for those who have sought to denigrate and ridicule public health protection; but fortunately, this is a phase that is rapidly being overcome by the massive amount of scientific evidence that now is being marshalled.

Concern about the health of painters has followed the general trend. Until recently, safety has been the predominant concern, and rightly so, because the accident rate in this trade is high. Now, concern about chronic health effects is mounting, but available data are limited, as recently was demonstrated during the International Symposium on Occupational Health Hazards Encountered in Surface Coating and Handling of Paints in the Construction Industry.* Fortunately, within the coming years we can expect a significant increase in knowledge as major studies are being completed. Today, however, more uncertainty prevails in this area than in most areas of occupational health, and the complexity of the field is almost insurmountable, especially in terms of designing adequate policies.

B. CURRENT REGULATIONS

By far the most sophisticated regulations of health in the painting trades are found in Scandinavia.[6] Basically similar, the Swedish,[7] Norwegian,[8] and Danish[9] regulations are based on the concept of air demand; that is, the amount of air required to dissipate hazardous vapors in specific paints to within the

*Held in Stockholm, Sweden, 3-5 October 1979.

range of existing TLV's. If more than one hazardous substance is found in the paint, the TLV-acceptable dissipation requirements for each substance are added together. Paints are then grouped into four categories according to the amount of air needed to dissipate hazardous vapors. For each category there are recommended health procedures. The Norwegian categories are listed in Table 1. Additional codes for especially hazardous substances may be added. This concept has recently been adopted by the Commission of the European Communities as well.[10]

The fundamental assumption of this regulation is thus the validity of TLV's. Although Norway has proposed to abandon the TLV approach, current regulations incorporate the TLV definition as established by the American Conference of Government Industrial Hygienists (ACGIH):

The Threshold Limit Value may be defined as that concentration of ambient pollution that almost all people can be exposed to throughout a full work shift every day for the duration of their working life without becoming discomforted and without receiving health damage.[11]

Rarely, however, do TLV's meet this definition. Study after study[12,13] reveal vast numbers of painters complaining of work-related discomforts. Given this situation, the first question that comes to mind is whether today's TLV's are adequate. At least for exposure to solvents, it is clear that painters often experience excess disease effects even when exposure is well within existing TLV's.[14,15] For this reason, it has been proposed that Time Weighed Average (TWA) limits should be set with strict 15-minute ceiling levels in an effort to control acute solvent intoxication, and subsequent chronic brain and nerve damage.[16]

It is necessary to take this question one step further and ask whether TLV's (or TWA's) are a valid concept for protecting the health of workers from chronic disease effects. Currently, this is a hotly debated issue. Dr. William Nicholson, of Mount Sinai School of Medicine, has carefully documented the reluctant step-by-step reduction in the TLV for asbestos in recent years as new scientific evidence documented disease effects at reduced levels of exposure.[17,18]

In Scandinavia, there seems to be general agreement that asbestos has no "safe" threshold level. Yet for other workplace hazards, including hazards to painters, the TLV concept is adhered to carefully.

Because of long latency periods, many synergistically interacting risks in the production of chronic diseases and varying biological responses to exposure, it does not seem to be adequate to rely on TLV levels for specific substances to protect workers' health. It is for these reasons that Norwegian policymakers in 1978 announced their intention to abandon the TLV approach.[19] Indeed, as long as we do not understand the biological mechanism by which an irreversible disease develops, the setting of "scientifically valid" thresholds of exposure is no more than a refined illusion.[20]

The central issue, then, is to propose a system of regulation that, as fully as possible, can protect workers from occupational health hazards, which in the absence of adequate data, we must assume produce irreversible, chronic diseases in many workers even at low levels of exposure.[21,22]

TABLE 1
Categorization of Paints' Hazards
Norway—1975

Category	Diffusion Level	Warning Label
1	Less than 100 m³/liter	"May be used without special concern about health or fire hazards, but ensure ventilation".
2	100-400 m³/liter	*"Dangerous to inhale in large quantities— Ensure good ventilation.* "Mechanical ventilation or appropriate respirator should be used when working in small or poorly ventilated spaces and when spray application is used".
3	400-800 m³/liter	*"Dangerous to inhale—Ensure good ventilation.* "Mechanical ventilation or appropriate respirator must be used always. When working in small or poorly ventilated spaces, and when spray application is being used, both mechanical ventilation and respirators must be used."
4	More than 800 m³/liter	*"Dangerous to inhale—Ensure good ventilation.* "Mechanical ventilation and appropriate respirators must be used always. When working in small or poorly ventilated spaces use fresh air respirator."

Source: Statens arbeidstilsyn. Arbeid med losemiddelholdig maling, lakk, lim o. l. Oslo 1975

C. THE CONCEPT OF NECESSARY RISK

As an alternative to the TLV approach, Sheldon Samuels is developing a regulatory theory that I think is both scientifically valid, and at the same time socially responsible. He has called this theory of risk reduction the Concept of Necessary Risk, on the basis of the following argument:

A citizen's *right to assume* that his government attempts to ban the unnecessary proliferation of carcinogens . . . . derives from his need for this protection. This and all rights arise from the *need* to preserve life . . . . The fact that we do not choose life is not an argument for the acceptance of risk. It is, instead, an argument for asking how to reduce risk. One possibility is in a concept of Necessary Risk, based on discoverable historical reality, that is, *the need to choose life utilizing not the form of precision identified with mere numbers but the substance of precision through the development of analytical tools consistent with the dialectics of democracy.* [23]

From this perspective, workers would be protected as fully as possible based on the following simple principles:

1. All unnecessary, irreversible risks should be eliminated.

2. Necessary substances posing an irreversible risk should be used under circumstances of adequate control only, including continuous monitoring of the worksite, and the workers, and with the full knowledge of those exposed.

3. New substances with lesser health risks than those presently in use should be sought to be developed.

4. New substances cannot be released without thorough pre-market health testing and appropriate government approval.

The ultimate goal of this policy would be to eliminate *existing* risks as fully as possible, and to *control* exposure to those risks that are deemed to be necessary as fully as possible. Thus, through a gradual process of product improvement, the intrinsic risks of the workplace would be reduced in the future. Through immediate engineering and personal control procedures, exposure levels at the present time would be minimized. As opposed to a futile discussion of acceptable environmental values, it would be recognized that improving workplace safety is a constant process. And instead of limiting the discussion of regulatory standards to the single biologic criterium included in the environment value, the debate would realistically include technological and economic feasibility, as well as means of control in the workplace and methods of enforcing the risk reduction.

The essence of the necessary risk concept is thus the extent to which current hazardous exposures serve vital social functions. At issue is the *controllability* of such exposures.

D. A CONCEPTUAL FRAMEWORK FOR
HEALTH PROTECTION
IN THE PAINTS TRADES

Ideally, protection of public health is targeted towards those risks that are known to be the greatest. Above all, the purpose of public policies and government programs must be to prioritize health needs, and to channel available resources from many disparate sources towards those needs that are the greatest. Thus effectiveness and efficiency in the use of limited resources can be maximized.

A policy is useful only to the extent that it can be implemented. Successful implementation depends on the means which policymakers have at their disposal. In all policy, there are three general categories of such means: (a) facilitating (or enabling) means, which are mainly fiscal ones and consist of incentives, such as subsidies, to direct flow of action towards the desired goals; (b) regulatory means, which consist of sanctions to prevent undesirable actions from taking place; and (c) education or information, which is considerably weaker than the two other means in that it has no direct material force. It is assumed, however, that an informed worker will be in a position to exert greater political leverage.

In order to establish priorities, and to determine the most effective and efficient means of implementation, available information must be systematized in a manner that correlates and contrasts all major factors involved. In Figure 1, this has been attempted for the paints trades. *What emerges is no simple matrix.*

First, painters cannot be grouped together in one category. The work of a painter of new automobiles bears little resemblance to that of a residential painter. The substances used by these two kinds of painters are largely different, as are the working environments and the modes of application. In this framework, painters are categorized according to the potential for control of risks. This is shown in the horizontal column.

The vertical column shows the relevant health-related factors. The risk of contracting occupational diseases is a product of:

1. *Toxicity.* The major toxic substances to which workers are exposed, and the health effects with which these are associated (e.g. cancer, central nervous system effects, dermatoses, etc.)

2. *Exposure.* The frequency (number of workdays exposed), duration (hours of exposure per day of exposure), and dose (or level) of the toxic substances to which the worker is exposed.

3. *Prevention Potential.* Substitutes for highly toxic substances, engineering capability to redesign the workplace to eliminate or reduce exposure levels, the use of personal protective equipment so that the individual worker can avoid exposure, biologic and environmental monitoring to minimize disease effect, and exposure levels and, the enforcement of effective regulations.

Ideally, on the basis of this matrix, policies could be developed to prioritize:
1. Protection of high risk groups;
2. Control of substances that are most hazardous;
3. Reduction of exposure patterns that are most hazardous;
4. Development of feasible prevention strategies.

E. PRIORITIES

1. Categories of Painters.

The necessary risk concept is centered on the question of controlling exposures. The question is, because the nature of the painter's work varies so much from setting to setting, how can we categorize the trade to develop a meaningful framework for prioritizing health protection? A search of available information revealed two existing categorization schemes, presented in Table 2. Both of these, however, were prepared for the purpose of research, and not for the purpose of control. A new categorization scheme had to be developed for this purpose.

If we return to the upper horizontal column in Figure 1, a scheme is proposed on the basis of controllability. The first delineation is between stationary and mobile painters. Stationary painters work in one spot while the objects that are to be painted come to them, usually on an assembly line. Examples are appliances, furniture and automobiles. Mobile painters work with stationary objects; hence, they must move both within the object (e.g., a ship, bridge, or house) and from object to object.

Stationary Painters. Two factors are of importance. First, a review of available information indicates that there is no technical reason for exposing stationary painters to hazardous substances. Excepting unanticipated events, a complete barrier between the painter and the painted objects can be feasibly erected at the present time, either through the development of self-contained areas for painting or through the use of personal protective equipment.[6]

Second, complete responsibility (and liability) for ensuring that such barriers are developed can be assigned to the industrial management. In this instance, the worksite is relatively permanent. Worker turnover is not high (unless conditions are already unsatisfactory), and routine health protection practices can be instituted, including clean locker rooms and eating areas.

Because of the feasibility of erecting complete barriers from exposure, the priority of policy in regard to stationary painters should be engineering, primarily, and personal protection, secondarily. Because reliance on personal protection and safe work habits invariably is less controllable than engineering of self-contained painting areas, engineering should ultimately receive the highest priority. Provided that the barriers from exposure are complete, use of hazardous materials is acceptable.

Mobile Painters. Mobile painters are divided into three categories according to the controllability criterium: (1) industrial painters; (2) structural painters;

FIGURE 1. A Conceptual Framework for Prioritizing Health Protection for Painters

Health Considerations	Categories of Painters				FEASIBLE PRIORITIES
	Stationary Painters	Mobile Painters			
		Industrial	Structural	Residential	
Toxicity –Major Substances –Major Health Effects					Elimination or Control of Most Hazardous Substances
Exposure –Frequency –Duration –Intensity (Dose)					Reduction of Most Hazardous Exposure Patterns
Prevention Potential –Substitution –Engineering Controls –Personal Protection –Environmental Monitoring –Biologic Monitoring –Enforcement					Implementation of Most Feasible Prevention Means
FEASIBLE PRIORITIES	Identification of Groups Workers at High Risk				

TABLE 2
Proposals for Categorizing Painters

Selikoff's Categories (1974)	Johns Hopkins' Categories (1979)
Industrial Painting	Appliance Painting
Structural Painting	Wood Furniture Painting
Residential Painting	Metal Furniture Painting
Sign Painters	Truck, Bus, Farm Equipment Painters
Floor Finishers	Railroad Painters
	Shipbuilding Painters
	Aircraft Painters
	Auto (original equipment) Painters
	Construction and Maintenance Painters

Sources: Selikoff, I.J. Investigation of Health Hazards in the Painting Trades. Mt. Sinai School of Medicine, New York, 1975.
NIOSH. Investigation of Health Hazards in the Painting Trades. Status Report for Contract No. 210-77-0096.

FIGURE 2. Controllability of Risks—Mobile Painters			
	Degree of Risk	**Degree of Controllability**	
		Fairly Good	**Poor**
	High	Industrial Painters	Structural Painters
	Moderate		Residential Painters

and (3) residential painters. The reason for this delineation is best illustrated by Figure 2. By correlating the hazardous exposures to the controllability of the working environment, it becomes apparent that industrial painters are exposed to high risks with good controllability potential; while structural painters are exposed to high risks with poor controllability potential; and residential painters are exposed to moderate risks with poor controllability potential.

Industrial Painters. Industrial painters are difficult to categorize. Materials used may be specialized, including the use of specialty paints containing antifouling agents and pesticides, which often are considered quite hazardous. Application is primarily by spray painting and dipping.

Because these painters are mobile, personal protective measures are commonly the major barrier against exposure. This may be a fairly adeqate method for protecting workers in the large industrial setting, where organizational stability and vast resources may combine to develop an adequate hygiene program. Often, however, even in massive plants, such as shipyards, the conditions of work are highly inadequate, and the health of painters is worse than what one might expect.

More severe is the situation in small industrial settings. Auto repair shops, small wood furniture shops, etc., are often dirty workplaces. In these settings the same person may clean, seal, glaze and sand the surface prior to also conducting the painting operation. The work area normally is ventilated poorly; and the worker poorly protected. Hence, the worker is exposed to a large number of risks that may interact, and the exposure is likely to be intense.[24] For these reasons, studies have revealed serious levels of behavioral toxicologic disease among car repair workers, even when levels of exposure are far below the TLV.[14]

Health protection in these areas must concentrate on all aspects of protection. First, and more immediate, workers must become aware of methods that can be used for personal protection given their exposure patterns. Second, an adequate system of enforcement and control must be established. Third, better engineering designs must be made available, through public subsidies to small businesses, if necessary. Fourth, substitute paints should be developed immediately. Fifth, workers should be monitored for long-term chronic effects.

Structural Painters. Structural painters work on bridges, early parts of construction, tanks, etc. The substances they work with are very toxic, including a number of pigments with carcinogenic and strong central nervous system effects. Paint application is often by the spray method, which enhances the hazardous environmental condition. The worksites are usually temporary and transient, with little possibility for adequate personal hygiene.

Residential Painters. The residential painters are probably exposed to the least hazardous agents, with the exception of asbestos from spackling, taping, and drywall materials. On the other hand, these painters (at least in the U.S.) are often not unionized, and both the trade and the industry is transient. The worksites are very temporary (average time for construction of a residential house in the U.S. is now around four months). Latex paints with low hazards are used, but the mode of application is increasingly spray painting, and ventilation is often poor.

Priorities for Structural and Residential Painters. Because of the transitory nature of the workplace and the painter's need to move about the worksite, controllability must emphasize the development of less hazardous, yet efficacious substances in paints, altered mechanisms of application, as well as improved personal protection.

Product Development—must particularly focus on pigments. (This will be discussed in greater detail in the next section).

Less Hazardous Application—would include eliminating spray painting where possible, and increase the use of brush and roll application, or the dipping of components prior to assembly. (The latter method, however, does not remove the hazards for welders during assembly and for maintenance and repair workers in the future).

Personal Protection—should include thorough education of painters about risks and the effectiveness of personal protection methods. Certification of painters could be a means to improve protection. Additionally, control of the worksite needs to be vastly improved, including the use of inspectors at each worksite who can monitor the ambient air quality and facilitate routine monitoring of workers. A central office should maintain detailed and continuous records on the work histories of individual painters. Since these painters tend to move from one employer to another, a government inspectorate should provide protection for non-union workers, while the unions may be a source of recordkeeping and control for their members.

The thought of inspectors at each worksite may seem excessive, but as the use of inspectors at each meat-packing and processing plant throughout the western world has demonstrated, it is a feasible mechanism.

2. Substances*

A breakdown of changes in major categories of substances in paints (see Table 3) reveals that while changes in overall consumption of substances has been erratic, a steady decline in the use of solvents, both in actual terms and as percentage of all substances used, has taken place in recent years. This is fortunate.

The decline in solvents use has been compensated for by an increase in the use of resins and pigments.

The composition of the substances used in some major areas of industrial painting are included in Table 4. A more detailed breakdown of paint components in a shipyard has been provided by Mount Sinai School of Medicine in New York.* Comparison of the lists in Table 4 and those in Appendix A reveals interesting differences. The shipyard painters appear to be exposed to more "severe" substances, particularly in the area of pigments. Also, benzene is found extensively in styrene/ethyl benzene

*Unless otherwise noted, sources of information for this section are references (25), (26), and (27).
*These are contained in Appendix A of this report.

solvents. However, all painters using hydrocarbon solvents can expect to be exposed to small amounts of benzene as a contaminant. The effect of this exposure is not known, but should be treated with extreme caution.

In applying the concept of necessary risk to the regulation of substances in paints, it is necessary to distinguish between old and new substances. New substances should be required to undergo scientifically valid tests for chronic health effects prior to marketing, and should not be marketed without government approval. Substances developed for paints directed at the general public should not pose known irreversible disease risks.

TABLE 3
U.S. Consumption of Raw Materials in Paints and Coatings
1973-1977
Billion of Pounds and % of Total

	1973		1974		1975		1976		1977	
	Total	%	Total	%	Total	%	Total	%	Total	%
Resins	2.27	23.3	2.24	24.6	2.04	24.5	2.20	25.4	2.27	26.1
Pigments	2.64	27.1	2.49	27.3	2.30	27.6	2.52	29.1	2.58	29.6
Solvents[1]	4.76	48.9	4.32	47.4	3.92	47.1	3.87	44.7	3.80	43.6
Additives[2]	0.06	.6	0.06	.6	0.06	.7	0.06	.7	0.06	.7
TOTAL	9.73	99.9	9.11	99.9	8.32	99.9	8.66	99.9	8.71	100

[1]Excludes water.
[2]Includes only surfactants, dispersing agents, drier cellulostics, thickeners, mildewcides, anti-skimming agents and anti-foaming agents

Source: Adapted from NPCA Data Bank, National Paint and Coating Association, as reported in O'Brien, D. et. al. Study Plan for an Evaluation of Engineering Control Technology for Spray Painting and Coating Operations. NIOSH, Cincinnati, December 1978

New substances directed at professional painters should be clearly marked with information about any ingredient that may pose an irreversible risk. Under any circumstances, and regardless of technical improvement, a new substance should be required to be significantly less hazardous than a substance that it replaces. Through the development of new substances and the elimination of old substances, the paint supply should progressively become less hazardous.

Substances that pose health hazards of an irreversible nature, and that are presently on the market, should be banned if they do not serve a technically vital function. A three year phase-out period could be announced for such substances to allow producers to seek alternatives. This would be the technology-forcing aspect of the regulation.

On the basis of available information, priorities for new product development can be formulated:

Substance	Auto Re-Finishing	Wood Furniture Production	Metal Furniture Production
TABLE 4			
Major Substances Used in Three Industrial Operations			
Percent of Total			
Solvents			
Xylene	32	15	18
Toluene	10	27	10
MEK	10	8	10
Ethyl Acetate	12		
Butyl Acetate	10		
N-Butyl Alcohol		12	5
Ethyl Alcohol		10	
Isopropyl Alcohol		10	
Aliphatic Hydrocarbons			36
Other	27	18	21
Resins			
Alkyd	47	42	
Acrylic	31		
Cellulostic	11	27	
Urethane	7		
Amino		8	
Other	4	23	
Pigments & Fillers			
Titanium Dioxide	33	26	60
Talc	14	25	6
Clay	12		
Baryles	12		7
Chrome	8		3
Iron Oxide	6		9
Lead	6		
Silica		18	
Calcium Carbonate		15	6
Other	9	8	9

Source: Same as Table 2.

Pigments and Fillers. Pigments that impart an esthetic quality only should be eliminated if they pose an irreversible disease effect, and a three year phase-out period could force producers to develop alternatives. Included in this category could be all lead-based pigments and chromium-based pigments. According to a recent industry-sponsored study, the main reasons for using lead-chromate paints are: "Firstly, certain stored industrial finishes are hard to obtain in the desired shade and fastness without the use of lead-chromate pigments. Secondly, it is generally accepted that small amount of lead drier are required for many air drying paints (*although published evidence for the exact function of the lead is somewhat limited*)."[6] Given these limited functions, and given these major caveats, a required phase-out should promote studies of the exact technical importance of lead and chromium and the development of viable substitutes. The reasons given in this quotation hardly justify the presence in paints of two sources of irreversible diseases (cancer and central nervous system disease).

It appears that mercury is being eliminated from paints, although its anti-fouling property could justify its use in rare circumstances and under extreme precaution.

To protect the residential painters, who are also exposed to asbestos, from additional risk, and to avoid a possible repetition of the environmental problem of the widespread presence of asbestos, both asbestos and talc should probably be eliminated.

The most commonly used pigment today, titanium dioxide, should be the subject of additional study. A common substance in titanium dioxide is amorphous silica, as well as minute trace elements of lead. Titanium dioxide has been reported to be a "respiratory irritant", the hazard of which remains unknown.[28] Although clinical studies of painters have not revealed detectable disease associated with exposure to titanium dioxide, they do indicate very high levels of respiratory system disease.[10]

Iron oxide has not been found to be hazardous to painters. It is thought that pure iron dust is not fibrogenic of the lung,[29] but that combined with other factors, it may be. Recent investigations of foundry workers indicate an elevated risk of lung cancer with a SMR of about 1.5.[30]

Solvents. The topic of solvents is of enormous complexity, and will not be covered in detail. However, it appears that aromatic hydrocarbon solvents (toluene, xylene, and styrene/ethyl benzene) are the most hazardous, and may be carcinogenic in addition to affecting the central nervous system. Benzene is a contaminant of these solvents. They enter the body through inhalation and skin contact. Blood cytogenic methods are too complex and of insufficient accuracy to be relied upon for early detection; and even with early detection, treatment and survival is unlikely. For these reasons, the German MAK commission has correctly concluded with regard to benzene;

On account of the proven carcinogenic (leukaemogenic) effects of benzene and the lack of quantitative measuring data in the low concentration ranges, it is not possible to establish an MAC which might be regarded to be without danger.[31]

Although these solvents may have important technical effects in paints, the health hazards involved dictate that the feasibility of eliminating from the market benzene and the products of which it is a contaminant, including commercial grade toluene, xylene, and styrene, should be studied.

Among ketones, methyl n-butyl ketone, either alone or together with methyl-ethyl ketone, appears to be most hazardous. Both are absorbed through the skin in addition to the vapors which are inhaled. The health effects may be extreme neurotoxic depression.[32] Great caution is required in the use of these substances.

Recently, potentially serious chronic health effects have been indicated for aliphatic hydrocarbons. N-hexane, an aliphatic hydrocarbon with metabolites that are very similar in structure to the toxic metabolites of methyl n-butyl ketone, may have a very similar peripheral neuropathy effect. They should, therefore, be treated with equal caution.

Other solvents of special concern are isobutanol and acetone, which have been classified by the EPA as potential high risk substances.

Hazardous solvents are often skin absorbent, and may be used improperly by painters for personal hygiene (cleaning of hands, etc.), at the completion of work. This practice warrants special precaution.

Resins. In the last two decades, a number of new substances have been developed. The resins are of particular importance. Epoxies, polyurethanes and acrylics are discussed here. These are used as binding agents in paints and floor coverings.

Epoxies are a group of hardeners introduced in the 1950's. They are very reactive and are thought to be carcinogenic in many cases. Particularly epichlorohydrin is considered to be a potential (respiratory) human carcinogen. Due to a wide range of positive findings in short term mutagenesis tests (Ames test), experimental animal tests and human studies, careful scrutiny should be given to the containued use of epoxy-based coatings.[33,34]

Polyurethanes were introduced into paints as a hardener in the 1950's. Isocyanates, the basis for polyurethane, were found to be extremely toxic in a volatile form, and a variety of alternatives have been developed. In varnishes especially, diisocyanates and polyisocyanates with low vapor pressure were developed, and the problem was thought to have been brought under control. Recent investigations, however, indicate that ". . . . up to now not sufficient attention was paid—to the reformation (of) diisocyanates as toxic vapors from polyfunctional hardeners and from varnishes at elevated temperatures."[35] Caution is thus justified.

Acrylics have been introduced in a variety of coatings in recent years. The acrylics encompass approximately 15 compounds, the most common of which is acrylonitrile.[36] Acrylonitrile is suspected of being a carcinogen on the basis of animal studies. Polymerization workers have been found to have excess rates of lung cancer and colon cancer. Skin, eye and nervous system effects (often non-specific) have been reported. Special concern is raised by the evidence that many acrylics are absorbed through the skin without detection. There is an urgent need to conduct appropriate research on the health effects of arcylics in coatings.

3. Exposure

Exposure to substances is most complex, and critical to the control of health risks. Three major variables are involved:
- *Frequency of exposure*—the number of times that the worker is exposed. A painter presumably is exposed every workday;
- *Duration of exposure*—the length of exposure for each time that the worker is exposed;
- *Intensity of exposure (or dose)*—includes: (1) the substance(s) that the worker is exposed to; (2) the mode of application, which promotes airborne particles or fumes (especially spray painting); (3) the adequacy of the ventilation systems, temperature control systems and other forms of engineering controls; (4) work patterns that may increase intake or absorption by the worker (e.g., increased ventilatory volume intake of solvent vapors due to physical stress); and (5) means of personal protection from skin contact, inhalation, or ingestion exposure.

Priorities for Control of Exposure. Among these alternatives, reducing the intensity of exposure must be the main objective in a program of health protection. Means for this have already been discussed. These include engineering, substitution, personal protection through the use of regulatory limitations and prohibition, facilitation of improved engineering through construction subsidies (for small businesses especially), education of workers (including the possibility of requiring certification for use of certain types of hazardous paints products), and careful inspection of workplaces and enforcement of regulations.

4. Prevention Potential

Critical to a program of occupational health protection is the determination of feasible strategies in the following areas:
- Substitution of Less Hazardous Substances
- Engineering Controls
- Personal Protection
- Environmental Monitoring
- Biological Monitoring
- Recordkeeping
- Enforcement

Substitution. Industry is constantly developing new products and substances. Until recently, concern about technical performance characteristics has predominated the development of new products, while health considerations have received little attention. There is reason to believe, that if the same effort were devoted to health concerns, the most hazardous substances in paints could be eliminated. Technology-forcing regulations (the stick), coupled with appropriate subsidies (the carrot), could create the environment conducive to such development. Table 5 summarizes the findings of this review with regard to current critical substances for which substitutes should be developed.

TABLE 5
Key Substances of Concern

Products that Should be Eliminated
Pigments and fillers: Lead, chromium, mercury, asbestos, talc
Solvents: Commercial grade aromatic hydrocarbons; methyl n-butyl ketone; n-hexane; isobutane; acetone
Resins: Epoxies

Products that Require In-depth Study as Used in Coatings
Pigments: Titanium dioxide
Resins: Polyurethanes; acrylics

Engineering Controls. The most effective measures for controlling exposure is the design and construction of workplaces that prevent direct contact between a worker and a hazard. Through improvements in *integral engineering*, self-contained areas are increasingly being developed for a number of occupational hazards. *Supplemental engineering* can promote much improved ventilation in the workplace.

Critical areas of development:

Stationary workers: Provide complete barriers to exposure (integral engineering);
Mobile workers: New methods of paint application, improved moveable ventilation systems (supplemental engineering); improved eating and locker room facilities.

Personal Protection. For mobile painters, personal protection is important under all circumstances at this time. Methods for personal protection have been developed, that if used correctly, can minimize exposure. Therefore, effective education of painters is of vast importance, and the certification of painters who are exposed to hazardous materials (especially those listed in Table 5) seems appropriate. Careful enforcement is needed to ensure that proper protective equipment is available to workers, and that the equipment is used properly.

Environmental Monitoring of Worksite. The control of workplace hazards requires information about the hazards to which workers are exposed. Continuous monitoring of the workplace environment is essential for this purpose.

Biological Monitoring. Biological monitoring of painters who work with hazardous substances (especially those listed in Table 5) is essential to the early detection of health effects. *However, biological monitoring can never be considered a primary source of health protection.* Few reliable tests have been developed. The tests cannot control for biological "outliers", e.g., individuals with abnormal responses to exposures. Early detection is meaningless if the detected disease is not curable. At the same time, biological monitoring may serve an innovative role in the prospective monitoring of workers who are exposed to new products which have not as yet been found to be hazardous in pre-market bioassay testing.

Recordkeeping. Systems of environmental monitoring and biological monitoring must include careful recordkeeping. Records for workers who are engaged at temporary worksites must be kept in a central location. Mechanisms to link the biological monitoring record of individual workers to the environmental monitoring records of the worker's occupational setting are required if either monitoring method is to be of long-term value for health protection. All systems of recordkeeping should give consideration to the most opportune use of the information collected for research purposes. This could promote the early detection of hazards and the identification of workers at risk of disease due to past hazardous exposures.

Enforcement. Improved enforcement is necessary for the protection of painters. Temporary worksites need to be inspected with special vigilance, and the assignment of resident inspectors may be appropriate.

F. CONCLUSIONS

In the course of history during the period marked by the market economy, economic competition within nations and among nations has proven to be a powerful vehicle for the promotion of technology. Unfortunately, the political economy of this period has not permitted much of that drive to be directed at public health. In general, the protection of workers' health has been incidental to the development of technologically desirable and less expensive products. Paints are no exception.

With policies based on the Concept of Necessary Risk, as discussed, and by establishing priorities in the areas proposed, we can expect to see a gradual elimination of hazardous products and improved workplaces in the paint trades. There is no reason why we should continue to find that the disease rates of painter are excessive if appropriate policies are implemented and enforced. By continuously prioritizing occupational health policies on the basis of a comprehensive examination of the work environments of painters, we should soon expect to see marked improvement in the health of painters.

ACKNOWLEDGMENTS

*Supported by contract No. B9-F94386 from U.S. Department of Labor. The author acknowledges the support of the Industrial Union Department, AFL-CIO and the Environmental Sciences Laboratory, Mt. Sinai School of Medicine.

REFERENCES

1. Hoel, DG et. al. Estimation of risk of irreversible, delayed toxicity. J Tox Env Health 1:133-151, 1975
2. Selikoff, IJ and EC Hammond. Multiple factors in environmental cancer. In Persons at High Risk of Cancer, edited by JF Fraumeni, Jr. Academic Press, New York, 1975
3. Bingham, E et. al. Multiple factors in carcinogenesis. Ann N Y Acad Sci 271:14-21, 1976
4. Lloyd, JW. Study of long-latent diseases in industrial populations. Wash Acad Sci 64:135-144, 1974
5. Armenien, HK and AM Lilienfeld. The distribution of incubation periods of neoplastic diseases. AM J Epid 99:92-100, 1974
6. O'Neill, LA. Health and Safety, Environmental Pollution and the Paints Industry: A Study Covering Legislation, Standards, Codes of Practice and Toxicology, with 206 References. Paints Research Association, Teddington (Middlesex) England, January 1977
7. Dufva, L. The influence of environmental factors on research and development of products for the decorative field in Sweden. Presented at the International Symposium on Occupational Health Hazards Encountered in the Surface Coating and Handling of Paints, Stockholm, October 3-5, 1979
8. Statens arbeidstilsynsdirektorat. Arbeid med lösemiddelholdig maling lakk, lim, o.l. Oslo, 1974
9. Direktoratet for Arbeidstilsynet. Malerarbejde: Anvisninger om foranstaltninger ved bygningsmaling. Köbenhavn, 1973
10. CEC. Council Directive of 7 November 1977 on the Approximation of the Laws, Regulations and Administrative Provisions of the Member States Regarding the Classification, Packaging and Labeling of Paints, Varnishes, Printing Inks, Adhesives and Similar Products. Official Journal of the European Communities, No. L 303/23, 28 November 1977
11. The ICGIH definition, as translated from the Norwegian regulation (see ref. 8)
12. Selikoff, IJ. Investigations of Health Hazards in the Painting Trades. Mount Sinai School of Medicine, New York, 1975

13. IFBWW. Occupational Health Hazards in the Painting Trades. Proceedings of the Conference 23-24 September 1976, Geneva, Switzerland. International Federation of Building and Wood Workers, 1977
14. Hanninen, H et. al. Behavioral effects of long-term exposure to a mixture of organic solvents. Scand J Work Envir. and Health 4:240-255, 1976
15. Seppalainen, AM et. al. Neurophysiological effects of long-term exposure to a mixture of organic solvents. Scand J. Work Environ and Health 4:304-314, 1978
16. Knave, B. Health hazards in the use of solvents. Presented at the IMF World Conference on Health and Safety in the Metal Industry, Oslo, Norway, 16-19 August, 1976
17. Nicholson, WJ. Occupational and environmental standards for asbestos and their relation to human disease. In Origins of Human Cancer, Book C, edited by HH Hyatt et. al. Cold Spring Harbor Laboratory, Cold Spring Harbor, 1977
18. Nicholson, WJ. Case study: Asbestos—the TLV approach. Ann N Y Acad Sci 271:152-169, 1976
19. Direktoratet for arbeidstilsynet. Administrative normer for forurensninger i arbeidsatmosfaere. Oslo, August 1978
20. Nelson, N. Comments on extrapolation of cancer response from high dose to low dose. Env Health Per 27:93-95, 1978
21. Rall, DP. Thresholds? Env Health Per 22:163-165, 1973
22. Rall, DP. Problems of low doses of carcinogens. J Wash Acad Sci 64:63-68, 1974
23. Samuels, SW. The role of scientific data in health decisions. Env Health Per 32:301, 1979
24. O'Brien, D et. al. Study plan for evaluation of engineering control technology for spray painting and coating operations. NIOSH, Cincinnati, December, 1978
25. Sax, NI. Dangerous Properties of Industrial Materials, 5th ed. Van Nostrand, Rheinhold, New York, 1979
26. Key, MM et. al. Occupational Diseases: A Guide to their Recognition. DHEW (NIOSH) Pub No 77-181, US Government Printing Office, Washington, 1977
27. Hamilton, A and HL Hardy. Industrial Toxicology, 3rd ed. Publishing Sciences Group, Acton, Massachusetts, 1974
28. Elo, R et. al. Pulmonary deposits of titanium dioxide in man. Arch Path 94:417-424, 1972
29. Teculescu, D and A Albu. Pulmonary function in workers inhaling iron dust. Int Arch Arbeitsmed 3:163-170, 1973
30. Egan, B et. al. A preliminary report of mortality patterns among foundry workers. In Dust and Disease, edited by R. Lemen and J. Dement. Pathotox, Park Forest, 1979

31. Senatecommission for the Examination of Hazardous Industrial Materials. Considerations Bearing on the Question of Safe Concentrations of Benzene in the Work Environment. Verlag Chemi, Weinheim, 1974

32. Spencer, PS and HH Schaumburg. Feline nervous systems responses to chronic intoxication with commercial grades of methyl n-butyl ketone, methyl isobutyl ketone and methyl ethyl ketone. Appl Pharm 37:301-311, 1976

33. NIOSH. Epichlorhydrin. Current Intelligence Bulletin no 30, October 12, 1978

34. Andersen, M et. al. Mutagenic action of aromatic epoxy resins. Nature 276:391-392, 1978

35. Seeman, J and U Wolke. Uber die Bildung toxischer Isocyanatdampfe bei der thermischen zersetzung von polyurethanlacken und ihren polyfunktionellen Hartern. Z bl Arbeitsmed 1976(1):2-9

36. Sjostrom, B. and B. Holmberg. A Toxicological Survey of Chemicals Used in the Swedish Plastics Industry. (In Swedish) National Board of Occupational Safety and Health, Stockholm, 1979

APPENDIX A*

Survey of Shipyard Paint Components

Reported Composition of 10 Thinners, Varnishes, and Solvents

(Out of a Total of 122 Compounds Described)

Hazardous ingredient reported	# reporting ingredient (N=10)	% of ingredient in total formulation range	average
Solvents:			
Aliphatic hydrocarbons			
mineral spirits	1		59.0
naphtha	2	500-100	75.0
Aromatic hydrocarbons			
xylene	8	45-100	82.4
Alcohols			
n-butyl alcohol	2	10-20	15.0
Ethers, Esters & Related Cmpds.			
butyl cellosolve	1		25.0
cellosolve	3	25-33	30.3
n-butyl acetate	1		45.0

---

*Prepared by Environmental Sciences Laboratory Mount Sinai School of Medicine, New York, NY 10029

Reported Composition of 18 Primers

(Out of a Total of 122 Compounds Described)

Hazardous ingredient reported	# reporting ingredient (N=18)	% of ingredient in total formulation	
		range	average
Solvents:			
Aliphatic hydrocarbons			
gilsonite	3	14.71-19.0	16.6
mineral spirits	8	3.0-37.0	17.7
naphtha	3	20.0-37.4	34.8
solvesso	1		29.0
Aromatic hydrocarbons	3	29.4-34.0	31.0
toluene	3	26.0-30.9	28.3
xylene	4	17.0-50.0	30.5
Ketones			
methyl ethyl ketone	1		25.0
methyl isobutyl ketone	3	0.7-28.0	9.8
Alcohols			
butanol	1		15.9
"denatured" alcohol	1		29.4
ethanol	2	0.3-0.5	0.4
isopropyl alcohol	1		60.5
n-propanol	1		5.0
Ethers, Esters & Related Cmpds.			
cellosolve acetate	1		6.0
Metals, Metallic Compounds			
lead oxide	4	9.0-49.8	29.5
magnesium silicate	1		16.0
zinc chromate	1		6.0

Reported Composition of 13 Epoxy, Resin Coatings

(Out of a Total of 122 Compounds Described)

Hazardous ingredient reported	# reporting ingredient (N=13)	% of ingredient in total formulation range	average
Solvents:			
Aliphatic hydrocarbons	1		2.9
mineral spirits	2	0.55-26.2	13.4
napthta	1		10.0
Aromatic hydrocarbons	1		18.4
ethyl benzene	1		5.0
toluene	1		5.0
xylene	9	0.35-14.8	8.0
Ketones			
methyl ethyl ketone	3	9.0-11.06	10.5
methyl isobutyl ketone	3	6.27-9.0	7.2
methyl n-butyl ketone	1		30.0
Alcohols			
butanol	7	0.44-6.5	2.6
ethanol	1		0.5
Ethers, Esters & Related Cmpds.			
cellosolve	3	0.88-3.65	2.7
cellosolve acetate	1		10.0
ethylene glycol monomethyl ether	1		6.9
Phenols	1		2.2
tri (dimethylamino) methyl phenol	2	1.76-2.0	1.9
Others			
ethylene diamine	1		0.35
nitrated aliphatic hydro-carbons	1		7.3
"ketimine" curing agent	1		11.4
Metals, Metallic Compounds			
cuprous oxide	1		70.0
lead chromate	1		5.0
lead oxide	1		45.0
zinc dust	1		70.0

Reported Composition of 16 Metal-Based Paints

(Out of a Total of 122 Compounds Described)

Hazardous ingredients reported	# reporting ingredient (N=16)	% of ingredient in total formulation	
		range	average
Solvents:			
Aliphatic hydrocarbons	1		1.9
coal tar	1		7.1
mineral spirits	12	4.56-48.0	25.6
Aromatic hydrocarbons	4	1.85-55.0	30.9
xylene	7	4.1-42.5	15.3
Alcohols			
butanol	2	2.9-5.0	3.9
ethanol	2	9.35-10.0	9.7
isopropanol	1		0.5
methyl isoamyl alcohol	1		1.8
Ethers, Esters & Related Cmpds.			
cellosolve	2	13.15-20.0	16.6
ethylene glycol monobutyl ether	1		10.0
Other			
epichlorohydrin	1		1.0
"red colorant"	1		1.0
Metals, Metallic Compounds			
lead chromate	1		28.4
lead oxide	1		30.0
litharge	1		0.3
silico lead chromate	3	5.0-6.0	5.3
titanium dioxide, calcium sulfate	3	30.0-40.0	33.3
zinc, metallic	3	60.0-80.0	70.0
Fillers, Inorganic Particles			
alkali silicate	1		5.0
magnesium silicate	1		11.0
mica (potassium aluminum silicate)	1		5.0

Resins:

phenolic ether resin	4	20.6-31.1	25.6
polyamides	2	13.3-14.0	13.7
urea resin	4	0.9-1.6	1.2

Reported Composition of 13 Anti-Fouling Paints and Coatings

(Out of a Total of 122 Compounds Described)

Hazardous ingredient reported	# reporting ingredient (N=13)	% of ingredient in total formulation range	average
Solvents:			
Aliphatic hydrocarbons			
"high solvency" hydro-			
carbons	1		20.0
mineral spirits	9	4.3-29.16	16.0
Aromatic hydrocarbons	2	15.4-58.0	36.7
xylene	7	4.9-15.0	9.4
Ketones			
methyl isoamyl ketone	1		5.0
methyl isobutyl ketone	1		8.1
Alcohols			
isopropanol	1		1.3
methanol	1		0.5
Phenols, Halogenated Phenols			
cresol	1		5.0
pentachlorophenol	1		5.0
Metals, Metallic Compounds			
copper "pigment"	1		31.2
cuprous oxide	9	20.0-70.0	32.1
mercuric oxide	1		0.4
tributyl tin flouride	1		7.8
tributyl tin oxide	2	1.8-5.0	3.4
zinc oxide	1		5.0
Resins:			
chlorinated rubber	1		14.0

Resins

chlorinated rubber	2		14.0
soya alkyd	3	30.0-35.0	33.3
phenol formaldehyde resin	1		7.0

Reported Composition of 52 Other Paints, Enamels and Boottoppings

---

(Out of a Total of 122 Compounds Described)

Hazardous ingredients reported	# reporting ingredient (N=52)	% of ingredient in total formulation	
		range	average

Solvents:

Aliphatic			
coal tar	3	11.5-24.8	16.5
mineral spirits	27	0.55-46.76	24.0
naphtha	6	15.0-60.0	38.8
petroleum spirits	1		20.0
Aromatic hydrocarbons	3	12.52-20.0	15.0
styrene	1		40.0
toluene	3	10.1-17.0	14.0
xylene	29	0.35-100	30.8
Ketones			
methyl isoamyl ketone	4	1.0-5.0	3.3
methyl isobutyl ketone	3	5.0-6.55	6.0
Alcohols			
butanol	6	0.44-10.0	2.8
ethanol	5	0.76-15.0	4.4
methanol	4	0.5-5.0	1.6
Ethers, Esters & Related Cmpds.			
cellosolve	5	0.88-15.0	4.6
cellosolve acetate	3	6.0-9.0	8.0
Phenols	3		2.2
tri (dimethylamino) methyl phenol	2		2.1
Other			
epichlorohydrin	6		1.0
epoxy resin coal tar	1		88.6
ethylene diamine	3		0.35
gelling agents	1		5.1
isobutyl isobutarate	1		6.0

"ketimine" curing agent	3		11.4
mixed amines and cresols	1		1.0
modified amines	1		10.0
2-nitro-propane	1		5.0
pentoxone	2		5.5
phenolic ether resin	1		20.7
Metals, Metallic Compounds			
litharge	1		0.4
molybdate orange	1		7.0
organotin antifouling cmpd.	1		8.0
Fillers, Inorganic Particles			
talc	1		45.6
Resins			
chlorinated rubber	3	12.0-18.0	14.0
amides	3	6.0-16.7	9.6

SECTION III

CHAPTER 9

WORK LOAD AND UPTAKE OF SOLVENTS IN TISSUES OF MAN

Irma Åstrand

*Unit for Work Physiology, Department of Occupational Health,
National Board of Occupational Safety and Health,
S-171 84 Solna, Sweden*

ABSTRACT

Subjects were exposed to different solvents in inspiratory air through a breathing valve both at rest and during physical exercise on a bicycle ergometer. The uptake in the organism was increased during physical work compared to at rest. The uptake was 5-6 times higher during heavy physical work compared to at rest for in blood easy soluble solvents whereas it was only doubled for more insoluble solvents. The uptake can be estimated from the quotient between the concentration in alveolar and inspiratory air and the pulmonary ventilation. The concentration in subcutaneous adipose tissue was higher in lean than in obese subjects. The half-life of methylene chloride in adipose tissue was about 8 hr and for styrene and xylene about 2-4 days.

INTRODUCTION

The uptake of solvents in the organism of man was measured in a series of experiments. This article summarizes some of the results which were obtained in exposure to toluene, industrial xylene, styrene, methylene chloride, methylchloroform, trichloroethylene, butanol and white spirit. In 1975 a review was written of results available at that time.[1] The present paper was presented in Stockholm to the International Symposium on Occupational Health Hazards Encountered in Surface Coating and Handling of Paints in the Construction Industry in October of 1979.

EXPERIMENTAL DESIGN, SUBJECTS AND METHODS

The same design of the uptake studies was used for all solvents. Young, healthy males were used as subjects. Each subject was exposed during both rest and exercise on a bicycle ergometer. Each period of exposure usually lasted 2 hr, 30 min at rest, 30 min at a work load of 50 W, 30 min at 100 W and another 30 min at 150 W. The solvent was given to the subject in the inspiratory air through a breathing valve. The concentration in the air was constant and it was usually kept around the Swedish threshold limit value of the solvent. By using the same work loads and exposure time for all solvents both pulmonary ventilation and blood circulation could be kept fairly constant during the 30 minute periods of exposure, meaning that the influence of the solubility in blood and tissues of different solvents could be studied.

The total uptake of the solvent was continuously measured with the Douglas' bag technique. The concentrations of solvents in air, in arterial and venous blood and in some instances in subcutaneous adipose tissue were measured by different gas chromatographic methods. Further details on the experimental design, analytical methods and error of the methods are given in the different part reports.[2,4-6,8-10]

RESULTS AND DISCUSSION

Uptake measurement. Examples of uptakes of methylene chloride, trichloroethylene, industrial xylene and styrene will be presented.

In Figure 1, the uptake of a subject exposed to methylene chloride is illustrated.[5] With increasing work loads, the pulmonary ventilation increased and thereby the given amount of the solvent. The pulmonary ventilation was about 6 times higher at 150 W than at rest. The amount taken up was about doubled during work compared to at rest, and of about the same size during the three cycling periods. During the whole period of exposure the percentage uptake declined from about 55% at rest, down to 25% at 150 W. The uptake in two hr in a group of 5 subjects was in mean 2.0 g, constituting about 31% of the inhaled amount of 6.5 g.

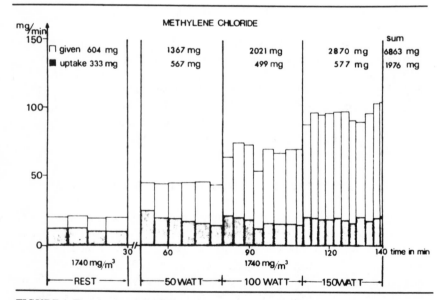

FIGURE 1 The amount of methylene chloride supplied and taken up in one subject during exposure. Exposure was performed with 1.740 mg/m³ of methylene chloride during rest and exercise at an intensity of 50, 100 and 150 W.[5] (Reprinted by Permission from CRC Press, Inc.)

In Figure 2, the uptake of a subject exposed to trichloroethylene is illustrated.[4] The uptake decreased during the whole period of exposure. It was about 55% during the first 30 minute period at rest and it succesively declined to about 6% during the last period at 150 W. It was almost zero at the end of exposure. After two hr, the concentration in the arterial blood of the subject was nearly in equilibrium with the concentration in the alveolar air. The subject was in good physical condition and fairly thin. The supplied amount in two hr to a group of 5 subjects was in mean 3.9 g. The uptake wa 1.4 g, constituting about 36% of the inhaled amount.

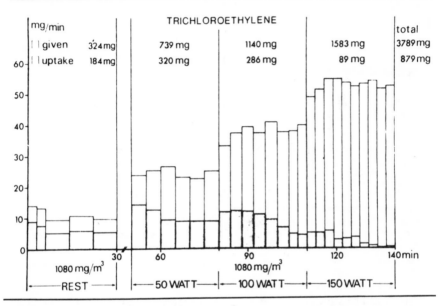

FIGURE 2 The amount of trichloroethylene supplied and taken up in one subject during exposure. Exposure was performed with 1.080 mg/m³ of trichloroethylene during rest and exercise at an intensity of 50, 100 and 150 W.[4] (Reprinted by Permission from CRC Press, Inc.)

In Figure 3, the mean uptake of a group of 6 subjects exposed to industrial xylene, a mixture of xylene isomers and ethylbenzene, is illustrated.[2] The uptake rose from about 80 mg during the rest period to about 390 mg during the last period of exercise. The percentage uptake was about 65% at rest and about 50% during 150 W. During the two hr exposure an average of 1.0 g or 59% was taken up of a supplied amount of about 1.7 g.

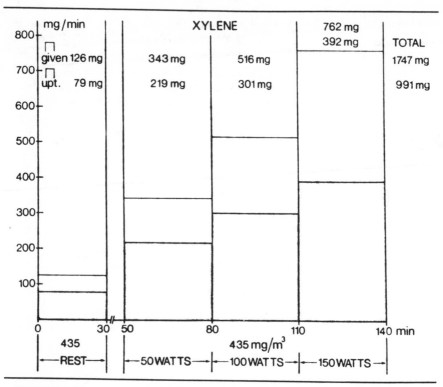

FIGURE 3 The mean amount of xylene supplied and taken up in six subjects during exposure. Exposure was performed with 435 mg/m³ of xylene during rest and exercise at an intensity of 50, 100 and 150 W.² (Reprinted by Permission from CRC Press, Inc.)

In Figure 4, the mean uptake of 7 subjects exposed to styrene is illustrated.¹⁰ The uptake was high during the whole exposure. About 63% or 0.5 g of the supplied amount of 0.8 g was taken up. The percentage uptake stayed high all the time. The uptake increased up to 5-6 fold during heavy physical work.

For all solvents, the uptake was at least doubled during physical work compared to at rest. For styrene and and xylene the uptake increased 5-6 fold during 150 W as compared to at rest, whereas the uptake for methylene chloride and trichloroethylene was only doubled. The explanation of these findings is that both styrene and xylene are easily soluble, whereas methylene chloride and trichloroethylene are comparatively insoluble in blood and probably also in the tissues. The more soluble the substance is the slower the equilibrium is reached between the concentration in blood and alveolar air and between tissues and blood. Especially, in the later part of the exposure the degree of metabolism also plays a role for the uptake.

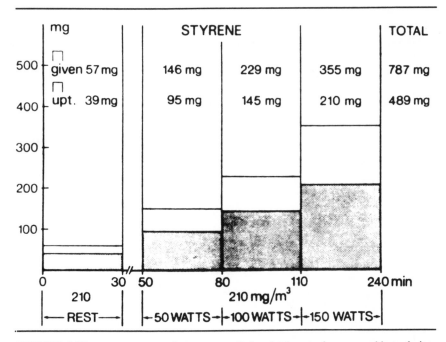

FIGURE 4 The mean amount of styrene supplied and taken up in seven subjects during exposure. Exposure was performed with 210 mg/m³ of styrene during rest and exercise at an intensity of 50, 100 and 150 W.[10] (Reprinted by Permission from CRC Press, Inc.)

Alveolar concentration. When the percentage uptake of the different solvents were related to the alveolar concentration over the inspiratory concentration, a linear correlation was observed.[1,2] (Fig. 5). The correlation between the two measures indicates that the uptake of a gas in the lungs approaches zero at the same time as the alveolar concentration approaches the concentration in inspiratory air, and that the percentage uptake increased when the concentration in alveolar air decreases.

The relationship between the two variables is natural. The fact that the uptake never reached 100% is due to the dead space ventilation. Accordingly, the amount taken up of a solvent may be estimated from the concentrations in the alveolar and inspiratory air and the pulmonary ventilation. This method of estimation was tried in the field with success.[7, 13] It should be noted, however, that the correlation only exists for solvent uptake in the lung capillaries.

The uptake in the lungs is distributed by the blood to the different tissues and organs. The higher the uptake per unit of time, the higher the concentration in the blood and the higher the concentration in different organs. This means that the concentration in vital organs varies despite a constant concentration in the ambient air. Therefore, what one would like to have is some sort of biological limit values as complements to the ordinary threshold limit values. Francesco Gamberale has worked together with our group. He made an extensive study of these problems.[12] He was able to show at which uptake a significant increase in simple reaction time can be demonstrated for the solvents discussed here. Such values can be used as biological limit values when discussing acute effects.[3]

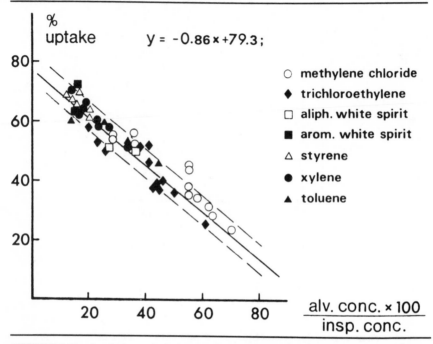

FIGURE 5 Uptake of solvents in the lungs as the percentage of the amount supplied in relation to the quotient between the concentrations of alveolar and inspired air. The uptake was measured during a 30 min period, and the corresponding alveolar concentration was based on three measurements during the last 10 min of the same period. In most cases each symbol represents a mean value of four to six subjects. The equation of the regression line was calculated on the basis of 46 such mean values. The number of subjects was: 14 for methylene chloride, 15—trichloroethylene, 4—white spirit, 7—styrene, 12—xylene and 7 for toluene (altogether about 60 subjects). The deviation from the line (SD) = ±5; = 0.949. (Reprinted by Permission from CRC Press, Inc.)

Content in subcutaneous adipose tissue. Jörgen Engström wrote his thesis on concentration of solvents in subcutaneous adipose tissue.[6] In Figure 6, the concentration of methylene chloride in subcutaneous adipose tissue after 1 hr exposure is given.[8] The concentration was, as a rule, higher in lean than in obese subjects. The concentration did not change significantly during the first 4 hr after exposure. The concentration was measured in 2 subjects the morning after the exposure and the estimated mean half-life of the concentrations in adipose tissue was 8 hr. The estimated mean amount of solvent retained in the total fat deposits of the body was greater in the obese subjects. The estimated total amount of methylene chloride in fat stores 4 hr after exposure was actually related to the estimated amount of the body fat expressed in per cent of body weight (Fig. 7).

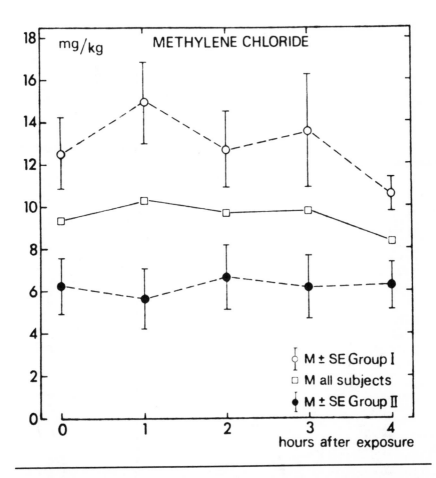

FIGURE 6 Concentration of methylene chloride in the subcutaneous adipose tissue of six slim subjects (group I) and of six obese subjects (group II) during 4 hr after exposure. Mean values of all 12 subjects are also given. Exposure was performed in 1 h with 2.600 mg/m³ of methylene chloride during exercise at an intensity of 50 W. The figure shows the mean values (M) and standard errors (SE) of the means.[8] (Reprinted by Permission from CRC Press, Inc.)

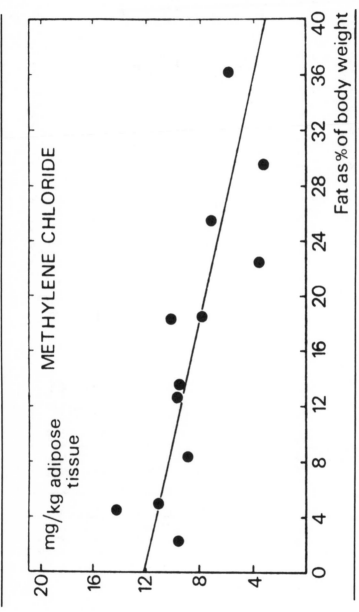

FIGURE 7 Concentration of methylene chloride in the subcutaneous adipose tissue of 12 subject 4 hr after exposure termination as related to the estimated fat % of body weight. Exposure was performed in 1 h with 2.600 mg/m^3 of methylene chloride during exercise at an intensity of 50 W.[8] (Reprinted by Permission from CRC Press, Inc.)

In Figure 8, the mean concentration in adipose tissue after 2 hr exposure to styrene is given.[10] The concentration did not change significantly during the first 24 hr. Four subjects were followed during a longer period of time after the exposure. In all 4 subjects, styrene was detectable for up to 13 days after exposure (Fig. 9). The half-life of the concentrations varied between 2 and 4 days.

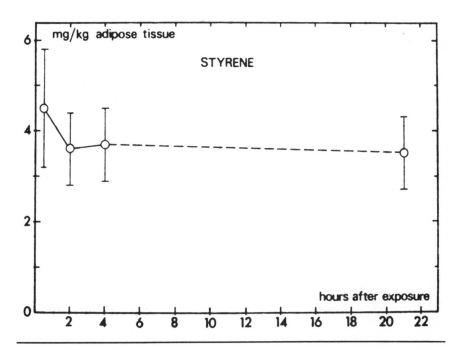

FIGURE 8 Concentration of styrene in the subcutaneous adipose tissue of 7 subjects in the first 24 hr after 2 hr exposure to about 210 mg/m³ in inspiratory air. Exposure was performed during rest and exercise at an intensity of 50, 100 and 150 W. Mean values and standard errors are given.[10] (Reprinted by Permission from CRC Press, Inc.)

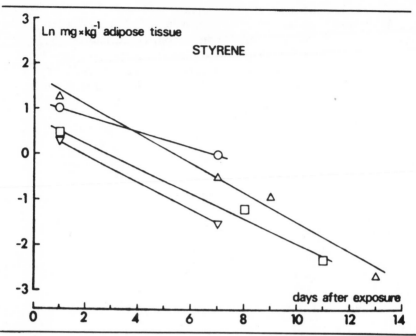

FIGURE 9 Concentration of styrene in the subcutaneous adipose tissue of 4 subjects from 1 to 13 d after 2 hr exposure to 210 mg/m³ in inspiratory air. Exposure was performed during rest and at an intensity of 50, 100 and 150 W.[10] (Reprinted by Permission from CRC Press, Inc.)

Figure 10 illustrates the concentrations in adipose tissue after exposure to industrial xylene.[9] There was no change in concentration during the first 24 hr. In a subsequent study the half-life of xylene in subcutaneous adipose tissue varied between 1 and 5 days.[11]

Because of poor blood perfusion of adipose tissue, only small amounts or around 10% of the total uptake were retained in adipose tissue. It is probably no risk of accumulation of methylene chloride in adipose tissue if the threshold limit values are not exceeded. However, because of the long half-life in adipose tissue of styrene and xylene, accumulation can be expected in occupational exposure. In three persons occupationally exposed to styrene, the mean concentration in adipose tissue was about 80 times higher than in ambient air.[7] There is reason to suspect that the elimination after industrial exposure takes much longer than two weeks.

What is the implication of an accumulation of a solvent in the adipose tissue? It means that as long as the solvent stays in the adipose tissue, it probably also stays in other fat-riche tissues. It also means that indirect exposure of internal organs such as the nervous system and the liver continues as long as the release from the adipose tissue continues. If the solvent and/or its metabolites are toxic the effect will of course be carried on during a longer time, and that is probably associated with a greater risk of development of diseases.

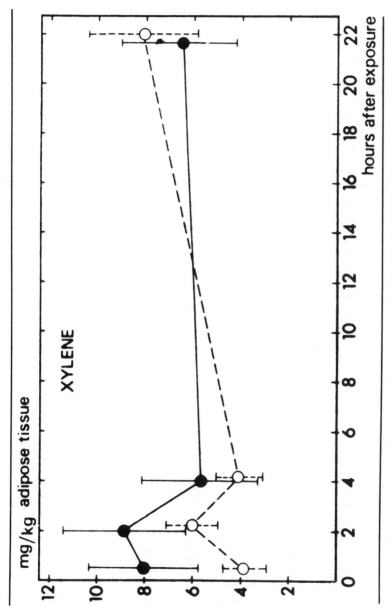

FIGURE 10 Concentration of xylene in the subcutaneous adipose tissue of subjects after exposure. • = mean and standard errors of 6 subjects exposed to 870mg/m^{3}in 2 hr during rest and exercise; o = mean and standard errors of 6 subjects exposed to 435 mg/m^3 in 2 hr during rest and exercise.[9] (Reprinted by Permission from CRC Press, Inc.)

REFERENCES

1. Åstrand I: Uptake of solvents in the blood and tissues of man. A review. *Scand J Work Environ & Health* 1:199, 1975.
2. Åstrand I, Engström J, Ovrum P: Exposure to xylene and ethylbenzene. I. Uptake, distribution and elimination in man. *Scand J Work Environ & Health* 4:185, 1978.
3. Åstrand I and Gamberale F: Effects on humans of solvents in the inspiratory air: A method for estimation of uptake. *Environmental Res* 15:1, 1978.
4. Åstrand I and Övrum P: Exposure to trichloroethylene: I. Uptake and distribution in man. *Scand J Work Environ & Health* 4:199, 1976.
5. Åstrand I, Övrum P and Carlsson A.: Exposure to methylene chloride: I. Its concentrations in alveolar air and blood during rest and exercise and its metabolism. *Scand J Work Environ & Health* 1:78, 1975.
6. Engström J: Organic solvents in human adipose tissue. *Arbete och Hälsa* 1978:22. Arbetarskyddsverket, Stockholm.
7. Engström J T, Åstrand I and Wigaeus E: Exposure to styrene in a polymerization plant. Uptake in the organism and concentration in subcutaneous adipose tissue. *Scand J Work Environ & Health* 4:324, 1978.
8. Engström J and Bjurström R: Exposure to methylene chloride. Content in subcutaneous adipose tissue. *Scand J Work Environ & Health* 3:215, 1977.
9. Engström J and Bjurström R: Exposure to xylene and ethylbenzene. II. Concentration in subcutaneous adipose tissue. *Scand J Work Environ & Health* 4:195, 1978.
10. Engström J, Bjurström R, Åstrand I. and Övrum P: Uptake, distribution and elimination of styrene in man. Concentration in subcutaneous tissue. *Scand J Work Environ & Health* 4:315, 1978.
11. Engström J and Riihimäki V: Distribution of m-xylene to subcutaneous adipose tissue in short-term experimental human exposure. *Scand J Work Environ & Health* 5:126, 1979.
12. Gamberale F: Behavioral effects of exposure to solvent vapors. Experimental and field studies. *Arbete och Hälsa* 1975:14. Arbetarskyddsverket, Stockholm.
13. Övrum P, Hultengren M and Lindqvist T: Exposure to toluene in a photogravure printing plant. Concentration in ambient air and uptake in the body. *Scand J Work Environ & Health* 4:327, 1978.

CHAPTER 10

RESPIRATORY SYMPTOMS DUE TO PAINT EXPOSURE

Kaye H. Kilburn, M.D.

Ralph Edington Professor of Medicine
University of Southern California
School of Medicine
Los Angeles, CA 90033

Painting creates fumes and aerosols, especially the latter. Brushing, rolling and spraying paint throw billions of fine particles into the atmosphere, that is, the breathing space of the painter and others working in the immediate air space.

Because there are systemic effects on the brain and other organs means that there has been pulmonary deposition and blood borne distribution to the brain, kidney, etc. This does not exclude cutaneous and gastrointestinal absorption in addition to the pulmonary route.

However, there are major effects upon the lung as a primary target organ. In contrast to the gastrointestinal tract and the skin, the lung is a cul-de-sac, a poorly cleared, blind ended retort in which poorly soluble materials have a long residence with opportunity to interact with cells. Viewed in this context, feeding experiments, even of long duration, have little possible relevancy to pulmonary deposition of particles. Secondary irritant effects in exposed workers, including symptoms such as coughing and wheezing, may represent only the tip of an iceberg when particles are present because chemicals adsorb to particles and elude from them after deposition in the lungs.

Experimental studies using formaldehyde and carbon black, both components of paint, have shown enhanced and different effects compared to those of formaldehyde alone (Kilburn and McKenzie, 1978). Even though pure carbon alone in massive doses had minimal effects in these studies, we have reinforced the conclusion that there are no truly *bland* particles. Macrophage recruitment and phagocytosis is the minimal result of particle exposure. When particle burdens delivered to the alveoli are high, macrophages may fill alveoli and seriously interfere with their function, reducing both diffusing capacity and total lung capacity.

Ideally, one should have data on well-studied populations of painters exposed to various component materials of paint—the pigments, vehicles and so-called inert carrier, in order to assess the effect of substance X on function Y. Not only are we without such data, except for a few materials such as asbestos, silica and, to some extent, lead chromate, but almost no general data exists on rates for the three principal pulmonary responses—asthma, chronic bronchitis and alveolar fibrosis in painters. To make matters more complex, painters are frequently exposed to asbestos, especially on shipboard or in dry wall taping, and to silica in sandblasting and paint texturing. Also, newer materials which resist wear and weathering, such as epoxy resins and polyurethanes, probably persist as particles in the lung.

Rather than reviewing existing data and concluding cleanly what health effects there are, because there are no data, one must instead extrapolate from general knowledge concerning pulmonary effects of molecules and of particles and especially of particles with adsorbed agents such as formaldehyde or trimillitic acid, for example, and sketch out some tentative probabilities and predictions. Although certain effects will be predicted, a plea must be made for cross-sectional and prospective surveys of pulmonary symptoms, function, and radiographic changes in painters. The magnitude of the problem is important, for there are at least 500,000 workers in the United States whose primary job is painting.

Results from a preliminary survey of nearly 1,000 painters attending a convention from four localities in the United States, showed a significantly increased prevalence of airway obstruction as determined by spirometry in 26% of non-smokers, and an associated increase of 9 to 12% in the symptoms of aireay hyersecretion as determined by the MRC questionnaire. (Selihoff, 1975) Thus, despite the lack of an opportunity to selectively relate exposure to pulmonary function, the results showing impairment and symptoms underline the problem of lung disease in painters and emphasize the need to segregate exposed painters in order to study changes in function with time to determine the relative contribution of solvents, particles, pigments, plastics and resins to annual decrements in pulmonary function.

Dr. Englund (see Chapter 12) showed that mortality ratios of Swedish painters are approximately 50% above those expected from chronic bronchitis and emphysema.

The case studies and general information about effects of individual agents to which painters are exposed permit a brief discussion of types of response which should be expected, and thus, allow for the design of studies designed to measure expected responses in painters. The four important types of responses are: first, asthmatic (reversible) airway narrowing; second, chronic airways obstruction with dyspnea (frequently called chronic bronchitis with disability); third, parenchymal, meaning eventually fibrotic lung disease, and lastly, lung cancer.

Asthmatic symptoms with chest tightness, wheezing and shortness of breath identify acute reversible stimulation of muscular medium sized airways. The mechanism may be direct, that is, pharmacological effect or via an immunological mechanism implying prior sensitization. Truly molecular amounts of an agent such as epoxy resin or polyurethane containing toluene diisocyanate (TDI), may be a sufficient stimulus and the response may be immediate, within minutes, delayed, taking several hours to reach a maximum, or at an interval coming on after 12 to 16 hours. Provocative testing is used which consists of mocking up or duplicating exposure in the laboratory and repeating tests of pulmonary function after first obtaining baseline measurements.

A great many substances of importance to painters have been shown to produce such responses including, most importantly, phthalic acid anhydride, isocyanates, trimellitic acid, triethylene tetramine (among 150 or more activators of epoxy resin), polyurethane varnish and formaldehyde.

Repeated exposure probably leads after a variable interval to hypersecretion and to metaplasia of goblet cells in small airways, mucuous obstruction and fibrous obliteration as has been chronicled in exposure to cotton dust and, in part, for toluene diisocyanate.

Chronic bronchitis, characterized by dyspnea, fixed obstruction of small airways and their scarring and obliteration is a most important industrial disease which painters are liable to develop. Undoubtedly, the combination of particle exposure and noxious chemicals produce injury by localizing damage to surfaces of airway cells. Such damage as by NO_2 or formaldehyde is measurably enhanced when gases are given simultaneously with particles. Adsorption to particles focuses and intensifies local damage, leading to leukocyte recruitment and to epithelial exfoliation. Double nuclei and apparent epithelial penetration of the basement membrane into the connective tissue also occur and may represent early premalignant changes.

Alveolar responses, the third major group of disorders begin with damage to the unique alveolar surface coating—surfactant. This disaturated lecithin (dipalmytol phosphotidylcholine) is greatly disordered or replaced by organic lipid solvents such as choloroform, methanol and by chaotropic agents such as cholesterol and other emulsifiers. The result is pulmonary edema. More protracted responses, especially to particles, include macrophage recruitment, proliferation and death, probably combined with stimulation or proliferation of macrophages to fill alveoli with cells and lipoprotein which is characterized as alveolar lipoproteinosis as originally described by Rosen, Castleman and Liebow in 1958.

The chronic alveolar responses associated with various insoluble or long lasting particles but most frequently silica is diffuse fibrosis. Although synthetic particles such as epoxy or polyurethane have not been associated as yet with this response or the previous one of lipoproteinosis, new methods of analytical electron microscopy will make it possible to detect these and fingerprint them including biopsies.

Neoplasia, specifically lung cancer, has been discussed, but chromate pigments appear to be an important potential cause of cancer in painters. Formaldehyde is currently under suspicion as a carcinogen. Many of the synthetic chemicals mentioned above have not run through a minimum (20 year) latent period, so cannot yet be assessed from human experience.

Perhaps from an orderly and sustained combination of studies the pulmonary responses of the painter's exposure can be fully appreciated. It may be predicted that the agents which have the very properties of resistance to weathering, adherence, color retention, etc., which make surface treatment most satisfactory will also have the strongest propensity for producing lung disease. If this prediction is true, then complete protection from exposure will become essential, for no lung residue will be permissible. Similar reaction molecules such as TDI, trimellitic acid, triethylene tetramine and formaldehyde which have strong sensitizing potential, I predict, can damage cells and probably lead to small airway disease? Properly designed studies of painters could resolve these questions and other important ones.

REFERENCES

1. Kilburn, K.H. and McKenzie, W.N. Leukocyte recruitment to airways by aldehyde-carbon combinations which mimic cigarette smoke. Lab. Invest. 38:134-142, 1978.
2. Rosen, S.H., Castleman, B. and Liebow, A.A. Pulmonary alveolar proteinosis. New Eng. J. Med. 258:1123-1142, 1958.
3. Selihoff, I.J. Investigations of Health Hazards in the Painting Trades. Mt. Sinai School of Medicine, New York, 1975.

CHAPTER 11

ORGANIC SOLVENTS AND KIDNEY FUNCTION

Alf Askergren, M.D.,
Bygghälsan, Sweden

INTRODUCTION

General

The large, heterogenous group of organic solvents contains several subgroups of chemically and often toxicologically related compounds. Solvents, most commonly commented upon from a nephrotoxicological point of view, are certain halogenated hydrocarbons (trichloroethylene, tetrachloroethylene, carbon tetrachloride, choloroform and other anesthetics), certain glycols (ethylene and diethylene glycol), alcohols (methyl, ethyl, isopropyl and butyl) and aromatic hydrocarbons (benzene, toluene, styrene and xylene).[1,2]

The following text will mainly deal with three of the most common aromatic hydrocarbons—styrene, toluene and xylene. *Styrene* plays an important role in the production of plastics and resins. In the manufacture of glass reinforced plastic products, such as plastic boats, it serves both as a solvent for the unsaturated polyester and as a monomer taking part in the polymerization.[3] *Toluene* and *xylene* are common solvents, thinners and cleaning agents, and both are current basic materials in the chemical industry. Xylene is one of the most commonly used solvents in paints and toluene is used in photogravure printing. Toluene may be the only solvent in a glue, otherwise it often predominates in a mixture including xylene, acetone, isopropyl alcohol, ethylacetate and trichloroethylene.[4,5]

Toxicologic aspects

Solvent vapours are almost solely taken up via the lungs.[6,7] The solubility of the substances in adipose tissues and the volume of body fats as well as the intensity of physical work and exposure influence uptake and distribution in the organism.[7,8]

Styrene is mainly metabolized in the liver. Via styrene oxide, it is excreted in the urine as mandelic acid and phenylglyoxylic acid. Approximately 80% of the *toluene* taken up by the organism is metabolized in the liver to benzoic acid, which is conjugated and excreted in the urine as hippuric acid.

The main metabolities of *xylene* are methylhippuric acids, which also are excreted in the urine.[9]

Our present knowledge of the nephrotoxicity of these solvents is chiefly based on animal experiments and on single human accidents. Conclusions cannot always be drawn from morphologic changes. When such changes are described, they usually are located to the tubules or the interstitial tissues,[10,11] but there are often simultaneous indications of glomerular function impairment.[12]

Laboratory results of exposing animals to varying amounts of these solvents are not quite in agreement: Fabre et al[13] found degenerative glomerular and tubular changes in the rabbit and the rat after 3000–5000 mg xylene/m[3] air for 40–130 days. Wolf et al[14] showed a significant weight increase in rat kidneys after up to 590 mg toluene/kg body weight/day or 667 mg styrene/kg/day orally for 6 months. In rats, guinea pigs and rabbits exposed to approximately 6000 mg styrene/m[3] air for 6 months a slight weight increase but no microscopic changes were found by Spencer et al.[15] Casts and granular debris were observed in the tubules of mice kidneys after 14000 to 32000 mg toluene/m[3] air for 7 hours.[16] Carpenter et al,[17,18] on the other hand, could find no significant micropathological differences between kidneys of non-exposed animals and animals exposed to toluene or xylene in amounts approximating those in the studies by Fabre et al.[13]

The consequences of glue sniffing are one of the most important documentations of solvent nephrotoxicity in humans. However here, too, the results are somewhat conflicting. Composition of glue, exposure level and methods of studying the effects on the kidneys vary. This explains at least some of the divergent results. O'Brien et al[5] presented a case of reversible anuria and hepatocellular damage after toluene sniffing. The most frequent findings, however, are proteinuria, hematuria and pyuria,[19,20] while other investigators have found no or only isolated cases showing such signs.[4,21] A few cases with impairment of renal tubular acidification have also been presented[22,23] and reversible renal insufficiency after heavy exposure to xylene has been described.[24]

There are few studies published on the effect on the human kidney of moderate or slight exposure to these three solvents. van Oettingen et al[25] exposed 3 persons to 50–800 ppm of toluene in several 8 hour experiments. They could find no signs of blood or albumin in the urine. Härkönen[26] found no cellular changes in the urine of 35 styrene exposed workers. Studies concerning mortality and morbidity in kidney and urinary tract disorders have not revealed any difference between subjects exposed to styrene and non-exposed subjects.[27-29] Greenburg et al[30] studied 106 aeroplane plant painters, 90% of whom were exposed to less than 700 ppm toluene. They found no differences in specific gravity, in urinary albumin content or in sediment from these painters, as compared to controls.

The possible association between organic solvent exposure and glomerular disease has been discussed in a number of papers: At least 13 cases are reported with the combination of glomerular disease and pulmonary alveolar hemorrhage. Nine of them had been exposed to petrol, the remainder to different mixtures.[31-34] Another 9 cases with glomerular disease and with an exposure to various organic solvents have been described.[31,35-38]

Four epidemiologic studies have been published, three of them suggesting a connection between glomerular disease and organic solvent exposure. Zimmerman et al[39] studied 63 adults with renal failure. There was a significantly greater exposure to toxic substances, above all hydrocarbon solvents, in those with proven or probable glomerulonephritis, compared to those with other renal disease. Lagrue et al[40] reached the same conclusions when comparing 108 non-systemic glomerulonephrites with 56 subjects with hypertension or urolithiasis. In a study by Ravnskov et al,[41] 50 biopsyverified glomerulonephrites were compared to 106 sex- and age-matched controls as to their subjective opinion on solvent exposure. Also in this study, there was a highly significant difference between the groups, the glomerulonephrites claiming greater exposure than did the controls. van der Laan,[42] on the other hand, found no difference in exposure between 50 cases of chronic glomerulonephritis and 50 controls.

The criteria used for selection of cases and controls vary between the four studies and are furthermore in some respects incompletely presented. This makes it difficult to compare the studies and to understand the causes for conflicting results.

To summarize, in spite of the present documentation, the possibility of an association between organic solvent exposure and glomerular disease is still to some extent an open question. Therefore a study has been performed to estimate the possible effect of moderate exposure to common organic solvents on some renal function measures, such as protein and cell excretion in the urine, renal concentrating ability and 51-Cr-EDTA clearance.

MATERIAL

Populations studied

Two groups of subjects were studied. One comprised 134 male workers who in their professions were exposed to organic solvents. Fifty-two were engaged in plastic boat manufacture (styrene exposure), 16 of which only took part in the protein study, 42 worked in photogravure printing plants (toluene exposure) and 40 in paint plants. These 40 were exposed to a mixture of solvents in which xylene and toluene dominated.

A control group to the exposed subjects comprised 48 male manual workers and employees. According to a carefully compiled occupational and leisure history, they had not been and were not being exposed to organic solvents to any substantial degree, i.e. not more than can be accepted as an average everyday exposure for any individual.

There were no differences between the controls and the exposed subjects with respect to case histories of previous renal and urinary tract disease.

TABLE 1.   Average age and body surface area (BSA) in the two groups.

Group	Average age	S.D.	Range	BSA	S.D.	n
Controls	47.6	14.0	20–64	1.93	0.11	48
Exposed	43.2	12.7	20–64	1.94	0.15	134

Exposure

The solvent vapour concentrations in ambient air were estimated in a gas chromatograph. The estimations of *styrene* concentrations showed that manual lamination below deck produced the highest amounts, up to 925 mg styrene/m³ air. During such jobs, masks were mostly used sporadically. As for the other jobs where the personal protection was limited or lacking, the concentrations of styrene in air varied between 100 and 300 mg/m³. The background levels were 20–100 mg/m³. The Swedich hygienic limit value for styrene is 170 mg/m³ air.

The 42 subjects exposed to *toluene* were employed in two photogravure printing plants with the same types of production.

The time weighted average values for toluene concentration varied between one-third and three times the Swedish hygienic limit value, being 300 mg/m³, with an average level of approximately the limit value. There was a rather equal distribution of toluene over the plants.

Xylene and toluene were dominating parts of the *solvent mixtures* present in the paint plants. All subjects were exposed to xylene and 36 to toluene. These two solvents constituted up to 80% of every individual's total exposure. Other solvents present in minor amounts were white spirit, methylene chloride, ethyl-, n-butyl- and isobutyl alcohol, methyl-, ethyl- and n-butyl acetate and methyl ethyl ketone.

A few of these subjects were for short periods exposed to concentrations 10 to 20 times exceeding the Swedish hygienic limit values for toluene, xylene, methylene chloride, n-butyl alcohol, isobutyl alcohol and n-butyl acetate, these values being 300, 350, 250, 150, 150 and 170 mg/m³ air respectively. The time weighted (one day) average concentrations however in no case exceeded these limit values and mostly varied between 10 and 50% of them.

METHODS

All participants were studied in the morning, fastening, after overnight fluid deprivation. The time schedule and the analyses performed are given in Figure 1. It was not known to the laboratory staff whether the subjects were controls or exposed.

	Time				
	21.00	Awake	8.30	10.00	11.00
Fluid deprivation					
Urine: Cells			x		x
Proteins			x		
Creatinine			x		
Osmolality	x		x	x	x
51-Cr-EDTA					
clearance					

FIGURE 1   Time schedule and performed analyses in the study of controls and exposed subjects.

If necessary the urine was buffered until pH was 6–7. It was frozen to $-20°C$ within two hours for later protein and creatinine analysis and estimation of osmolality. All urine analyses were performed within 14 days.

Protein pattern was determined by means of agarose electrophoresis.[43]

For *albumin* analysis urine was concentrated 100 times (Minicon[R] B 15, Amicon, USA) and the amount was determined by means of radial immuno-diffusion (Partigen[R], Behringwerke-Hoechst, West Germany) according to Mancini et al.[44]

Beta-2-microglobulin was determined by radioimmunoassay (Phadebas[R], beta-2-micro test, Pharmacia, Sweden).

Creatinine concentration in urine was determined kinetically with the alkaline picrate reaction (initial reaction rate).

Microscopy was performed on 1 μl uncentifuged urine, in a Bürker Counting Chamber within 30 minutes of micturition.

Osmolality was determined by measuring the freezing point depression (osmometer, type M, Knauer, West Germany).

Glomerular filtration rate was estimated by calculating the *5-Cr-EDTA clearance* with conventional methods.[45]

RESULTS

Protein excretion

Figure 2 and Tables 2 and 3 summarize the estimations of albumin and beta-2-microglobulin in the urine.[46] The average excretion of albumin was significantly higher in the exposed group ($p = 0.003$) and especially so in the subgroup exposed to styrene. There was also a significantly higher number of subjects in the control group excreting less than 1 mg albumin/1 urine ($p = 0.004$). There were more subjects in the exposed group than in the control group ($p = 0.054$) exreting more than 10 mg albumin/1, especially so in the subgroup exposed to styrene ($p = 0.022$).

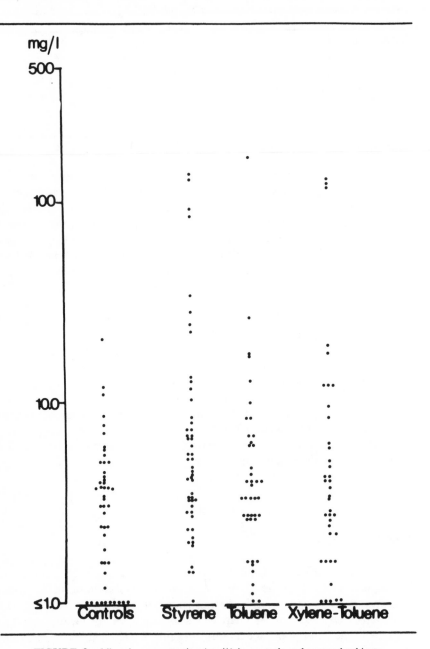

FIGURE 2   Albumin concentration (mg/1) in controls and exposed subjects.

No correlation could be found between the estimated degrees of exposure and the excreted amounts of albumin.

TABLE 2. Albumin concentration in urine (mg/1).

	Mean	S.D.	Median	Range	n
Controls	3.7	3.7	3.1	0– 21	48
Exposed	15.8	47.0	3.9	0–440	133
Styrene	23.0	65.9	4.9	1–440	52
Toluene	8.7	24.9	3.2	0–163	42
Xylene-toluene	13.7	32.2	3.2	0–130	39

TABLE 3. Concentration of beta-2-microglobulin in urine (μg/1) and number of subjects excreting more than 400 μg/1 and more than 1000 μg/1.

	Mean	S.D.	Median	Range	n	No. of subjects excreting >400 μg/1	>1000 μg/1
Controls	259	393	134	0–1840	48	6	3
Exposed	185	380	132	3–4200	134	11	2

There was no statistically significant difference in beta-2-microglobulin excretion between the exposed group and the control group. Nor was there any such difference in number of subjects excreting more than 400 or 1000 μg/1.

The agarose electrophoresis findings were in agreement with the patterns presented above. One exposed subject excreted immunoglobulins and transferrin, another 5 exposed subjects also excreted transferrin.

Cell excretion

Figures 3 and 4 and Tables 4 and 5 illustrate the results of estimating the cell excretion in the exposed subjects and the controls.[47] As shown, there was a significant difference in both erythrocyte and leucocyte excretion between the exposed and the controls. This significance was present both when the excretion was expressed as cells per liter urine and as cell excretion per minute.

No certain association could be found between the different exposure levels in the three subgroups of exposed and the amount of cells excreted.

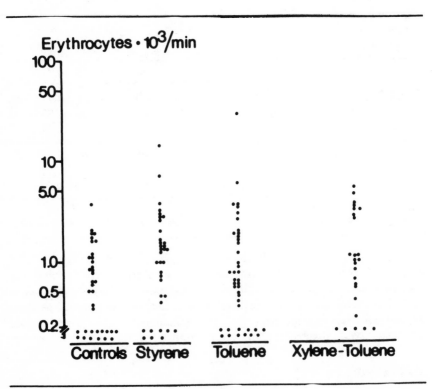

FIGURE 3   Erythrocyte excretion (• 10³/min) in urine from controls and exposed subjects.

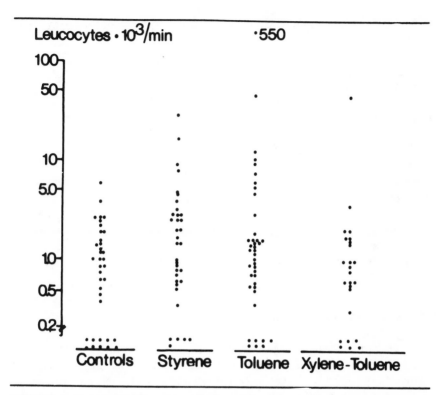

FIGURE 4 Leucocyte excretion ($\bullet 10^3$/min) in urine from controls and exposed subjects.

TABLE 4. Excretion of cells ($\bullet 10^6$/1) in the morning urine of the groups of controls and exposed subjects .

	Mean	S.D.	Range	p	n
Erythrocytes					
Controls	1.5	1.8	0– 8	0.011	39
Exposed	3.1	5.6	0–50		99
Leucocytes					
Controls	2.6	2.9	0–12	0.014	39
Exposed[a]	6.2	13.7	0–84		97

[a]Two subjects, excreting 600 and 1000 $\bullet 10^6$ cells/1 are excluded.

TABLE 5. Cells excreted per minute in urine of the groups of controls and exposed subjects.

	Mean	S.D.	Range	p	n
Erythrocytes					
Controls	734	830	0– 3520	0.012	36
Exposed	1644	3368	0–27500		101
Styrene	1693	2558	0–14000		35
Toluene	1689	4324	0–27500		41
Xylene-toluene	1502	1529	0– 5220		25
Leucocytes					
Controls	1144	1240	0– 5830	0.010	36
Exposed[a]	3078	7255	0–46500		100
Styrene	3095	5420	0–27700		35
Toluene[a]	3306	7632	0–46500		40
Xylene-toluene	2689	8976	0–45600		25

[a]One subject excreting 550000/min is excluded.

Renal concentrating ability

No difference in average renal concentrating ability after 11–14 hours of fluid deprivation could be found between the controls and the exposed subjects, as shown in Figure 5. Nor was there any difference in number of subjects not reaching 800 mosmol/kg (6 controls and 18 exposed).[48]

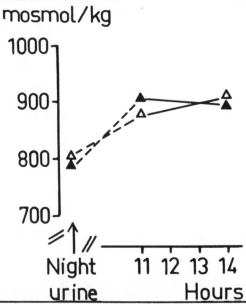

FIGURE 5   Urinary osmolality in 48 controls (Δ) and 118 exposed (▲) after 11–14 hours of fluid deprivation.

Glomerular filtration rate

There was no significant difference between the groups in average 51-Cr-EDTA clearance (Table 6).[49]

TABLE 6. 51-Cr-EDTA clearance in controls and exposed subjects.

| | ml/min/1.73 m² BSA | | | |
	Mean	S.D.	p	n
Controls	100.7	15.9		48
			0.103	
Exposed	107.5	14.7		107

DISCUSSION

Protein excretion

Modern techniques such as the concentration of urine protein, electrophoresis and immunochemical methods have made possible detailed identification and specific quantification of different proteins.[50]

Increased glomerular permeability mainly results in increased excretion of proteins the size of albumin or greater.[51] Augmented excretion of low-molecular size proteins such as beta-2-microglobulin, on the other hand, rather is the result of pathological conditions affecting the mechanism for absorption, transport and degradation of proteins in the proximal tubules.[52] Accordingly, glomerular and tubular as well as mixed forms of proteinuria are present in a great number of pathological conditions in various parts of the kidneys and urinary tract.[53,54]

Many of the different studies on the amounts of proteins normally present in the urine, have been performed under such conditions that they cannot be considered comparable as to factors, known to influence the protein concentration. Such factors are age,[55] pH,[56] posture,[57] exercise,[58] stress,[59] fever[60] and urine flow.[61] The question of near normal excretion of proteins therefore is difficult to penetrate in the literature.

The selection of the subjects to the control and the exposed groups in the present study make the two groups comparable in terms of factors mentioned above, and there was no difference in urine flow between the groups.

Cell excretion

There are several physical and physiological factors affecting quantification of cells in the urine, such as centrifugation procedure, unstandardized amounts of urine studied, urinary pH and osmolality.[62-64] Studying uncentrifuged urine in a counting chamber within about one hour after micturition eliminates these factors.

In 1928, Dukes[65] suggested the estimation of cell excretion in the urine by means of studying uncentrifuged urine in a counting chamber, the investigation being performed on morning specimens. A few studies on urinary cell excretion have since then been performed using a quantitative method and a standardized protocol, hence making them comparable. The results of this study's control group agree well with those of three such studies.[66-68] As mentioned earlier, the selection of subjects to the groups of controls and exposed makes the groups comparable as to physical and physiological factors which also can have affected the urinary cell content. Therefore, the results indicate that there is a difference in cell excretion between subjects who in their profession are exposed to organic solvents, compared to non-exposed persons, even if the difference is small.

Renal concentrating ability

Determining the osmolality of urine by estimation of the freezing point depression of the urine is a convenient method with high reproducibility for estimation of the urinary concentration of solutes.[69] By now, 800 mosmol/kg is a commonly used limit value for discriminating between normal and abnormal concentrating ability after a fluid deprivation peroid.

Several drugs and other substances can impair the renal concentrating ability.[11] Even if the clinical picture after heavy exposure to solvents includes impaired tubular function, no study seems to have established any decrease in concentrating ability after moderate to light exposure. Nor has this study indicated any such impairment of tubular function, when comparing the effect of fluid deprivation on the average osmolality of the exposed and the controls, or when comparing the number of subjects not reaching 800 mosmol/kg in the two groups.

Glomerular filtration rate

51-Cr-EDTA is by now a well documented tracer and one of the most frequently used for measuring GFR by a singe bolus injection.[70,71]

As pointed out by Rahill,[72] the GFR may be normal early in renal disease in spite of increased glomerular filtration of plasma proteins. The fact that this study has not shown any difference in glomerular filtration rate, as estimated by 51-Cr-EDTA clearance, between controls and exposed, does not exclude the possibility of early changes in the renal structures of subjects exposed to organic solvents.

SUMMARY

There are convincing indications of an association between massive exposure to organic solvents and renal damage. The possible connection between even slight to moderate exposure and glomerular disease of immunologic character has lately been suggested in epidemiologic studies. However, this has hardly been verified in conventional toxicologic investigations with estimation of glomerular filtration rate, protein excretion and cell excretion as well as microscopic examination of renal tissue. No epidemiologic study with simultaneous estimation of renal function at various levels of the nephron seems as yet to have been presented.

51-Cr-EDTA clearance, albumin and beta-2-microglobulin excretion, renal concentrating ability and cell excretion were measured in 134 subjects exposed to various organic solvents, in particular styrene, toluene and xylene, and 48 controls. The controls were chosen to be comparable to the exposed except for solvent exposure conditions. There was no difference between the groups as to beta-2-microglobulin excretion, concentrating ability or glomerular filtration rate. The exposed subjects on an average excreted more albumin and more cells than the controls. The electrophoresis of urine also indicated a glomerular pattern of protein excretion in a few of the exposed by the presence of transferrin and immunoglobulins. No such proteins could be found by electrophoresis in the urine of the controls.

There was no certain association between the increased albumin and cell excretion and the exposure levels. Whether this is explained by the rough differentiation into exposure levels masking a dose-response relationship, or by the absence of such a relationship, cannot be concluded from the present study.

The findings of increased albumin excretion suggest enhanced permeability of the glomerular basement membrane. The cause of increased cell excretion is not known, but may be related to the same mechanism as the increased albumin excretion. The changes in the kidneys were not advanced enough to show any impairment in GFR or proximal or distal tubular function.

These results, indicating a difference in the glomerular structure in subjects exposed to moderate amounts of common organic solvents, are of interest both as a sign of possible direct nephrotoxic effect and in the light of recent observations of an association between solvent exposure and glomerular disease of possible immunologic character. Whether organic solvents predispose to glomerulonephritis and/or cause an aggravation of pre-existing such disease, or cause a direct toxic injury to the kidneys remains to be shown.

ACKNOWLEDGEMENTS

This study was supported by the Swedish Work Environment Fund.

REFERENCES

1. Browning, E.: Toxicity and metabolism of industrial solvents. Elsevier Publishing Company, 1965.
2. Cornish, H.H.: In: Casarett and Doull's Toxicology. Macmillan Publishing Company Inc., New York. 2nd ed.: 468–496, 1980.
3. Tossavainen, A.: Scand. J. Work. Environ. Health 4, suppl. 2: 7–13, 1978.
4. Massengale, O.N., Glaser, H.H., Le Lievre, R.E., Dodds, J.B. & Klock, M.E.: N. Engl. J. Med. 269: 1340–1344, 1963.
5. O'Brien, E.T., Yeoman, W.B. & Hobby, J.A.E.: Br. Med. J. 2: 29–30, 1971.
6. Riihimäki, V. & Pfäffli, P.: Scand. J. Work. Environ. Health 4: 73–85, 1978.
7. Riihimäki, V., Pfäffli, P., Savolainen, K. & Pekari, K.: Scand. J. Work. Environ. Health 5: 217–231, 1979.
8. Åstrand, I.: Scand. J. Work. Environ. Health 1: 199–218, 1975.
9. Toftgård, R. & Gustavsson J.-Å.: Scand. J. Work. Environ. Health 6: 1–8, 1980.
10. Ehrenreich, T.: Am. Clin. Lab. Sci. 7: 6–16, 1977.
11. Schreiner, G.E.: In: Edwards (ed.): Drugs affecting kidney function and metabolism. Progr. Biochem. Pharmacol. Vol. 7. S. Karger, Basel: 248–284, 1972.
12. Foulkes, E.C. & Hammond, P.B.: In: Casarett & Doull (eds.): Toxicology. Macmillan Publishing Company Inc., New York: 190–200, 1975.
13. Fabre, R., Truhaut, R. & Laham, S.: Arch. Mal. Prof. 21: 301–313, 1960.
14. Wolf, M.A., Rowe, V.K., McCollister, D.D., Hollingsworth, R.L. & Oyen, F.: Arch. Ind. Health 14: 387–398, 1956.
15. Spencer, H.C., Irish, D.D., Adams, E.M. & Rowe, V.K.: J. Indust. Hyg. Toxicol. 24: 295–301, 1942.
16. Svirbely, J.L., Dunn, R.C. & van Oettingen, W.F.: J. Indust. Hyg. 25: 366–373, 1943.
17. Carpenter, C.P., Geary Jr, D.L., Myers, R.C., Nachreiner, D.J., Sullivan, L.J. & King, J.M.: Toxicol. Appl. Pharmacol. 36: 473–490, 1976.
18. Carpenter, C.P., Kinkead, E.R., Geary Jr, D.L., Sullivan, L.J. & King, J.M.: Toxicol. Appl. Pharmacol. 33: 543–558, 1975.
19. Press, E. & Done, A.K.: Pediatrics 39: 611–622, 1967.
20. Sokol, J. & Robinson, J.L.: West Med. 4: 192–194, 1963.
21. Barman, M.L., Sigel, N.B., Beedle, D.B. & Larson, R.K.: Calif. Med. 100: 19–22, 1964.
22. Fischman, C.M. & Oster, J.R.: JAMA 241: 1713–1715, 1979.

23. Taher, S.M., Anderson, R.J., McCartney, R., Popovtzer, M.M. & Schrier, R.W.: N. Engl. J. Med. 290: 765–768, 1974.
24. Morley, R., Eccleston, D.W., Douglas, C.P., Greville, W.E.J., Scott, D.J. & Anderson, J.: Br. Med. J. 3: 442–443, 1970.
25. von Oettingen, W.F., Neal, P.A. & Donahue, D.D.: JAMA 118: 579–584, 1942.
26. Härkönen, H.: Int. Arch. Occup. Environ. Health 40: 231–239, 1977.
27. Frentzel-Beyme, R., Thiess, A.M. & Wieland, R.: Scand. J. Work. Environ. Health 4, suppl. 2: 231–239, 1978.
28. Lorimer, W.V., Lilis, R., Fishbein, A., Daum, S., Anderson, H., Wolff, M.S. & Selikoff, I.J.: Scand. J. Work. Environ. Health 4, suppl. 2: 220–226, 1978.
29. Thiess, A.M. & Friedheim, M.: Scand. J. Work. Environ. Health 4, suppl. 2: 203–214, 1978.
30. Greenburg, L., Mayers, M.R., Heimann, H. & Moskowitz, S.: JAMA 118: 573–578, 1942.
31. Beirne, G.J. & Brennan, J.T.: Arch. Environ. Health 25: 365–369, 1972.
32. Heale, W.F., Matthiesson, A.M. & Niall, J.F.: Med. J. Aust. 2: 355–357, 1969.
33. Klavis, G. & Drommer, W.: Arch. Toxikol. 26: 40–55, 1970.
34. Sprecace, G.A.: Am. Rev. Respir. Dis. 88: 330–337, 1963.
35. Cagnoli, L., Casanova, S., Pasquali, S., Donini, U. & Zucchelli, P.: Br. Med. J. 1: 1068–1069, 1980.
36. Ehrenreich, T., Yunis, S.L. & Churg, J.: Environ. Res. 14: 35–45, 1977.
37. Kleinknecht, D., Morel-Maroger, L., Callard, P., Adhémar, J.-P. & Mahieu, P.: Arch. Intern. Med. 140: 230–232, 1980.
38. von Schéele, C., Althoff, P., Kempi, V. & Schelin, U.: Acta Med. Scand. 200: 427–429, 1976.
39. Zimmerman, S.W., Goehler, K. & Beirne, G.J.: Lancet II: 199–201, 1975.
40. Lagrue, G., Kamalodine, T., Hirbec, G., Bernandin, J.F., Guerrero, J. & Zhepova, F.: Nouv. Presse Med. 6: 3609–3613, 1977.
41. Ravnskov, U., Forsberg, B. & Skerfving, S.: Acta Med. Scand. 205: 575–579, 1979.
42. van der Laan, G.: Int. Arch. Occup. Environ. Health 47: 1–8, 1980.
43. Johansson, B.G.: Scand. J. Clin. Lab. Invest. 29, suppl. 124: 7–19, 1972.
44. Mancini, G., Carbonara, A.O. & Heremans, J.F.: Immunochemistry 2: 235–254, 1965.
45. Bröchner-Mortensen, J. & Rödbro, P.: Scand. J. Clin. Lab. Invest. 36: 35–43, 1976.
46. Askergren, A., Allgén, L.-G., Karlsson, C., Lundberg, I. & Nyberg, E.: Acta Med. Scand. 209: 479–483, 1981.
47. Askergren, A.: Acta Med. Scand. 210: 103–106, 1981.

48. Askergren, A., Allgén, L.-G. & Bergström, J.: Acta Med. Scand. 209: 485–488, 1981.
49. Askergren, A., Brandt, R., Gullquist, R., Silk, B. & Strandell, T.: Acta Med. Scand. 210: 373–376, 1981.
50. Berggård, I.: In: Manuel, Revillard & Betuel (eds.): Proteins in normal and pathological urine. S. Karger, Basel (Switzerland), New York: 7–19, 1970.
51. Pollak, V.E. & Pesce, A.J.: Perspect. Nephrol. Hypertens. 6: 155–174, 1977.
52. Carone, F.A.: Ann. Clin. Lab. Sci. 8: 287–294, 1978.
53. Jensen, H.: Dan. Med. Bull. 19: 89–98, 1972.
54. Revillard, J.P., Fris, D., Salle, B., Blanc, N. & Traeger, J.: In: Manuel, Revillard & Betuel (eds.): Proteins in normal and pathological urine. S. Karger, Basel (Switzerland), New York: 188–197, 1970.
55. Mahurkar, S.D., Dunea, G., Pillay, V.K.G., Levine, H. & Gandhi, V.: Br. Med. J. 1: 712–714, 1975.
56. Wibell, L.: Pathol. Biol. 26: 295–301, 1978.
57. Robinson, R.R. & Glenn, W.G.: J. Lab. Clin. Med. 64: 717–723, 1964.
58. Poortmans, J. & Jeanloz, R.W.: J. Clin. Invest. 47: 386–393, 1968.
59. Mogensen, C.E., Gjøde, P. & Christensen, C.K.: Lancet I: 774–775, 1979.
60. Hemmingsen, L. & Skaarup, P.: Acta Med. Scand. 201: 359–364, 1977.
61. Berggård, I. & Risinger, C.: Acta Soc. Med.Upsaliensis 66: 217–229, 1961.
62. Gadeholt, H.: Br. Med. J. 1: 1547–1549, 1964.
63. Gadeholt, H.: Acta Med. Scand. 183: 49–54, 1968.
64. Stansfeld, J.M.: Arch. Dis. Child. 37: 257–262, 1962.
65. Dukes, C.: Br. Med. J. 1: 391–393, 1928.
66. Gadeholt, H.: Acta Med. Scand. 184: 323–331, 1968.
67. Gerhardt, W., Stein, G., Feustel, E. & Schuster, B.: Dtsch. Gesundheitswesen 25: 1345–1348, 1970.
68. Krecke, H.J. & Schütterle, G.: Dtsch. Arch. Klin. Med. 207: 118–139, 1961.
69. Bevan, D.R.: Anaesthesia 33: 794–800, 1978.
70. Chantler, C., Garnett, E.S., Parson, V. & Veall, N.: Clin. Sci. 37: 169–180, 1969.
71. Favre, H.R. & Wing, A.J.: Br. Med. J. 1: 84–86, 1968.
72. Rahill, W.J.: In: Rubin & Barratt (eds.): Pediatric Nephrology. The Williams & Wilkins Company, Baltimore: 10–40, 1975.

CHAPTER 12

CANCER INCIDENCE AND MORTALITY AMONG SWEDISH PAINTERS

Göran Engholm and Anders Englund
BYGGHÄLSAN, the Construction Industry's Organization for Working Environment, Safety and Health, Stockholm, Sweden

INTRODUCTION

Studies on the distribution of causes of death in different occupations including painters has been reported from the US[1] and UK.[2] However, there is no evidence of any European study conducted on painters in the construction industry.

In Sweden, Bygghälsan has studied the causes of death and incidence of cancer among some of the occupations involved in the construction trade. Painters, plumbers/pipefitters and insulators are among them. Two different sets of exposure information (Union membership files and Population Census data) and two different sets of outcome information (both cancer incidence data and mortality data) have been utilized for this report.

MATERIALS AND METHODS

Data-set 1

Retrospectively, a cohort was defined from the membership files of the Swedish Painter's Union. This file, computerized from 1966, has been the starting point of followup. A person was considered to have entered his cohort (and starting his exposure as a painter) when he became a member of the Union. The time of entry into the study varies among the subjects.

The cohort has been followed-up through 1974 with regard to mortality. This has been done through record-linkage with the annual registries on causes of death kept by the Swedish Central Bureau of Statistics. Underlying and contributory causes of death has been registered by that Bureau according to the International Classification of Diseases. The 7th revision (ICD VII) has been used through 1967 and ICD VIII (8th revision) from 1968 onwards.

With respect to cancer incidence a similar record-linkage procedure with the Swedish Cancer Registry has been applied. That Registry, which was established 1958 and is being kept by the National Board of Health, has permitted followup throuh 1971. The cancer registry receives reports from both the clinically treating physician and the pathologist. The cancer registry uses the International Classification of Diseases 7th revision (ICD VII). Reporting is compulsary by law to both these registries. Similar operations have been made simultaneously for cohorts of plumbers/pipe fitters and insulators of which the former will be used for some comparisons.

In the mentioned record linking, use is made of the unique ten-digit identification number, which is assigned to every person living in the country. For details of this procedure reference is given to a previous report.[3] Corrections have been made of all incorrect identifications and only 50 out of the total 50,800 persons (1 0/00) could not be traced through this procedure. The length of followup varies among the subjects and the maximum duration extends from 1966 through 1971 for cancer incidence and through 1971 for cancer incidence and through 1974 for mortality. The number of persons studied—for both painters and plumbers/pipe fitters—and the corresponding number of person-years observed is shown in Table 1.

The expected numbers of deaths in different causes and of reported number of cancer cases in different sites have been calculated using 5-year-interval age groups and calendar year specific numbers of person-years of observation. The corresponding male mortality and incidence figures applied have been obtained from official Swedish reports on Causes of Death[4] and on Cancer Incidence.[5] The relation between observed and expected has been calculated as an SMR-value. In addition to SMR-values for the painters, such SMR-values—for the benefit of comparison—have been calculated also for the plumbers/pipe fitters.

The significance of the differences between observed and expected numbers has been calculated as one-tailed p-values. These have been computed using exact Poisson-distributions when the expected number is less than ten. Otherwise the Poisson-distributions has been approximated with the normal distribution.

Data-set 2

The Swedish Cancer Registry contains data on some 350,000 cancer cases reported between 1960 and 1973. The Population Census of 1960 contains data on approximately 7.5 million inhabitants of Sweden that year. These two registries have been linked together using the previously described unique ten-digit personal identification number. In the census file 99% of the cancer cases reported have been identified, which leaves only 1% "lost to followup".[6] Among the available data in the census file, those on occupation and industrial trade have been studied and reported.[7, 8]

The observed numbers of cases for each site in different occupations have been compared with what could be expected considering the number of persons and the age and sex distribution in those occupations and the corresponding incidence figures for the total Swedish population. The difference between observed and expected has been calculated as an SMR-value. The ratios of observed to expected number of cancer cases in different

TABLE 1

SIZE OF PAINTER AND PLUMBER/PIPE FITTER COHORTS AND NUMBER OF PERSON YEARS OF OBSERVATION

	Painters	Plumbers
Number of workers	30,580	18,521
Person years of follow-up mortality	238,026	145,646
D:o cancer morbidity	152,754	90,194

TABLE 2

MORTALITY AND CANCER INCIDENCE AMONG PAINTERS AND PLUMBERS IN THE SWEDISH HOUSE BUILDING TRADE

	Occupation	Source of inf	OBS	EXP	SMR	p-value
Total mortality	Painters	SCB	2740	2690	1.02	0.18
	Plumbers	SCB	836	939	0.89	0.0005
Cancer incidence	Painters	Cancer Reg.	647	590	1.09	0.01
	Plumbers	Cancer Reg.	236	211	1.12	0.06

sites for those 25,805 who were painters exclusively in the house painting trade have been calculated.

RESULTS

Data-set 1

Total mortality and cancer incidence in the two cohorts of painters and plumbers/pipe fitters is recorded in Table 2. The expected and observed numbers of deaths is very close for the painters. For the plumbers/pipe fitters the SMR is 0.89. Table 3 shows the mortaltiy for painters in some non-malignant causes of death and its relation to the number of years since the painter entered the cohort. The mortality in cardio-vascular (ICD VIII 410-14) or cerebrovascular (ICD VIII 420-38) diseases do not deviate from the expected. There is a 40-50% excess mortality for chronic bronchitis, emphysema and asthma (ICD VIII 490-93) and the longer the time since entrance into the cohort the higher the excess. Such an increase over time is not found for deaths in non-malignant diseases of the esophagus and stomach although the general level of excess deaths is of a similar magnitude (SMR 140-145).

Although mortality statistics could be expected to be an insensitive indicator of mental disorders, the mortality figures for some diagnositc entities in that category will be presented and contrasted with a somatic diagnosis (Table 4). Within the total cohort there is a 40-50% excess in deaths due to psychosis, neurosis, and alcoholism as the major entity. The excess mortality is more pronounced among those who have belonged to the cohort for less than 20 years. In spite of that excess mortality shown for alcoholism there is no excess mortality for cirrhosis of the liver among the painters. Although the suicide rate among the painters is not remarkably elevated compared to that of the general Swedish male population, it is almost twice as high as that of a comparable industrial worker group in the construction trade—the plumbers/pipe fitters (Table 5).

The morality in some malignant diseases is shown in Table 6 and incidence in total and selected cancer sites in Table 7. The excess in total cancer incidence increases as the number of years since entrance into the painter cohort is increasing (Table 7). For those who entered 20 years ago or more, the excess in cancer incidence is 15%. The excess mortality (Table 6) and incidence (Table 7) in esophageal cancer is about two-fold and of the same magnitude regardless of time since entrance into the cohort. There is no equivalent excess in neither mortality nor incidence for cancer of the stomach. There is a two-fold excess in incidence of cancer in the gall bladder and the intrahepatic bile ducts (ICD VII 155.1 - 155.8) with a possible increase over the years (Table 7). The excess mortality in cancer of the peritoneum (ICD VIII 158) increase from two-fold for the total cohort to an SMR of 346 for those who became painters more than 25 years ago (Table 6). The observed

TABLE 3

MORTALITY AMONG PAINTERS IN CERTAIN NON-MALIGNANT DISEASES

Site (ICD VIII)		Years since entry into Union					
		≥0	≥5	≥10	≥15	≥20	≥25
Ischemic heart disease (410, 414)	OBS	947	933	886	809	739	666
	SMR	1.00	1.00	1.00	0.98	0.98	1.00
	P	0.4602	0.4602	0.5000	0.2743	0.3085	1.5000
Cerebrovascular disease (430, 438)	OBS	213	207	195	179	170	152
	SMR	0.94	0.94	0.92	0.90	0.94	0.95
	P	0.2119	0.1841	0.1357	0.0968	0.2119	0.2743
Chronic bronchitis, emphysema, asthma (490-93)	OBS	63	63	60	58	55	49
	SMR	1.41	1.45	1.46	1.52	1.59	1.62
	P	0.0035+++	0.0019+++	0.0019+++	0.0007++++	0.0003++++	0.0003++++
Esophagus, stomach (530-37)	OBS	52	52	47	43	38	34
	SMR	1.45	1.50	1.44	1.42	1.39	1.43
	P	0.0035+++	0.0019+++	0.0062+++	0.0107++	0.0228++	0.0179++

TABLE 4

MORTALITY AMONG PAINTERS IN CERTAIN NON-MALIGNANT DISEASES

Site (ICD VIII)		Years since entry into Union					
		≥0	≥5	≥10	≥15	≥20	≥25
Psychosis (290-99)	OBS	6	6	5	5	5	4
	SMR	1.44	1.48	1.30	1.39	1.53	1.39
	P	0.2417	0.2229	0.3434	0.2913	0.2316	0.3275
Neurosis (300-309)	OBS	21	18	13	13	10	6
	SMR	1.53	1.45	1.20	1.43	1.36	1.08
	P	0.0287+	0.0668	0.2743	0.1309	0.2097	0.4804
Alcoholism (303)	OBS	20	17	12	12	9	6
	SMR	1.50	1.41	1.14	1.36	1.25	1.11
	P	0.0359+	0.0808	0.3446	0.1822	0.2937	0.4540
Cirrhosis/liver (571)	OBS	38	35	29	25	23	17
	SMR	1.04	1.02	0.94	0.91	0.99	0.91
	P	0.4207	0.4602	0.3821	0.3446	0.5000	0.3446

TABLE 5

SUICIDE AMONG PAINTERS AND PLUMBERS (ICD VII)

| | | ≥0 | ≥5 | Years since entry into Union | | | | |
				≥10	≥15	≥20	≥25
Painters	OBS	112	96	75	60	49	39
	SMR	1.19	1.20	1.15	1.15	1.18	1.25
	P	0.0359+	0.0446+	0.1151	0.1587	0.1357	0.0808
Plumbers	OBS	38	18				
	SMR	0.64	0.69				
	P	0.0035---	0.0668				

TABLE 6

MORTALITY AMONG PAINTERS IN CERTAIN MALIGNANT DISEASES

Site (ICD VIII)		Years since entry into Union					
		≥0	≥5	≥10	≥15	≥20	≥25
Cancer of Esophagus (150)	OBS	24	22	21	21	20	18
	SMR	1.95	1.84	1.86	2.01	2.12	2.20
	P	0.0005++++	0.0019+++	0.0026+++	0.0007++++	0.0018+++	0.0021+++
Cancer of stomach (151)	OBS	80	78	74	72	64	54
	SMR	1.06	1.06	1.06	1.12	1.10	1.06
	P	0.3085	0.3085	0.3085	0.1841	0.2420	0.3446
Cancer of peritoneum abdominal wall (158)	OBS	5	5	5	5	5	5
	SMR	1.98	2.11	2.30	2.57	2.92	3.46
	P	0.1115	0.0919	0.0697	0.0477+	0.0304+	0.0160++
Cancer of lung (162)	OBS	124	118	114	103	92	80
	SMR	1.27	1.25	1.28	1.27	1.26	1.27
	P	0.0047+++	0.0082+++	0.00047+++	0.0107++	0.0139++	0.0179++

TABLE 7

CANCER INCIDENCE AMONG PAINTERS

Site (ICD VII)		≥0	≥5	Years since entry into Union ≥10	≥15	≥20	≥25
Total all causes	OBS	647	617	583	531	480	410
	SMR	1.09	1.10	1.12	1.12	1.14	1.15
	P	0.0107++	0.0107++	0.0035+++	0.0047+++	0.0035+++	0.0035+++
Esophagus (150)	OBS	17	16	15	15	13	11
	SMR	2.15	2.10	2.10	2.29	2.20	2.18
	P	0.0034+++	0.0053+++	0.0069+++	0.0032+++	0.0078+++	0.0144++
Intrahepatic bile ducts (155.1, 155.2, 155.3, 155.8)	OBS	12	12	11	11	9	9
	SMR	2.00	2.07	2.01	2.18	1.98	2.30
	P	0.0202++	0.0158++	0.0246++	0.0144++	0.0424+	0.0187++
Larynx (161)	OBS	14	13	12	11	8	7
	SMR	1.77	1.74	1.73	1.76	1.45	1.54
	P	0.0315+	0.0422+	0.0499+	0.0533	0.1931	0.1749
Lung (162, 162.0, 162.1 162.8, 163, 163.2)	OBS	81	75	74	66	58	51
	SMR	1.28	1.24	1.31	1.28	1.26	1.32
	P	0.0139++	0.0359+	0.0139++	0.0228++	0.0446+	0.0287

number of laryngeal cancers is close to twice the expected number (Table 7). Both mortality (Table 6) and incidence (Table 7) of lung cancer show a 25-30% excess with no difference between the total cohort and those who joined the trade long ago.

There is an almost two-fold excess in chronic lymphatic leukemia in both the mortality (17 observed cases) and cancer incidence computations. However, the number of observed cases is substantially lower than expected in other categories of leukemias.

Data-set 2

Among the 25,805 persons who in 1960 reported to the population census that they were painters in the building trade, 1,511 cancer cases were reported between 1960-73. The observed numbers of cases in selected sites and the corresponding SMR-values are listed in Table 8. There is a 65% excess in cancer of the esophagus (ICD VII 150) but not excess in stomach cancer (ICD VII 151). The number of cases of cancer in the gall-bladder and intrahepatic bile ducts (ICD VII 155.1-8) is twice the expected and the excess in pleural tumors (ICD VII 162.2) is almost three-fold. However, the number of cases is minimal in both these sites. The ratios between observed and expected incidence of cancer in the larynx (ICD VII 161) and lung (ICD VII 162.0-1) are both 1.30.

DISCUSSION

The observed number of deaths among the painters is identical to the expected, which is an unusual finding in an industrial cohort. The ratio of observed to expected of 0.89 observed among plumbers/pipe fitters is more consistent with the "healthy worker effect" frequently observed in such cohorts.

The total number of reported cancer cases exceeds the expected among those painters who entered into the cohort sufficiently long ago to allow for the "latency period" of two—three decades frequently seen in occupational cancers. Entrance into the cohort always took place around the age of 20 when the vocational training was completed but the time of exposure to the critical substance might be another one. We do not even know which factor in the work environment—if any—should be suspected to be responsible for one disease or another. Neither exposure time, dose, nor latency time can thus be estimated. The differences in cancer incidence between urban and rural areas in Sweden and such due to certain personal habits like smoking are not likely to have influenced the findings, as we have no reason to believe that the painters differ to an important degree from the typical Swedish male population in these respects.

The almost two-fold excess in chronic lymphatic leukemias observed in both the mortality and the incidence part of the study must be interpreted with caution. Although there is no reason to assume a higher degree of misclassification

TABLE 8

CANCER INCIDENCE IN TWO
POPULATION CENSUS OCCUPATIONS

	Diagnosis (ICD VII)	SMR	Number of cases
Painters/house-	150	1.67	32
building only	151	1.12	159
(1511)	155.1-8	2.02	7
	161	1.32	26
	162.0-1	1.30	202
	162.2	2.68	5
Shoemakers	150	1.71	7
(579)	155.0	1.87	8
	155.1-8	2.88	3
	204	1.56	21

Total number of cancer cases in that occupational category within brackets.

TABLE 9

CANCER OF THE ESOPHAGUS (ICD VII 150) IN SOME
SELECTED OCCUPATIONS

Occupational category	Total number of cancer cases	Number of esophageal cancer cases	SMR
Painters	1511	32	1.67
Shoes and boot industry	579	7	1.71
Rubber industry	843	11	1.90
Chemical industry	408	11	2.12
Petrol station	238	5	2.13

for painters than for other occupations, the observation that the number of cases with other forms of leukemia is substantially lower than expected, should be kept in mind.

The disease pattern among the painters is similar with the use of both mortality and morbidity data, and both in the specific Union-defined cohort and in the Census-defined one. The pattern is similar to the one seen in alcoholics. There is an excess in deaths in that particular diagnosis among the painters. However, this is a rarely used diagnosis in Sweden and as the number of cases diagnosed as cirrhosis of the liver does not show a similar deviation from the expected, suspiciousness can be raised as to the validity of the diagnosis alcoholism. In another Swedish study it has been observed that early retirement due to mental disorders was more common among painters than controls and that no difference existed between the groups at the age of military service.[9] The finding that the suicide rate among the painters, although rather close to the national average, is almost double to that of the comparable construction worker group of plumbers adds to this suspiciousness. It should be noted that the Swedish suicide statistics is likely to be most accurate. One possibility that has to be ruled out is that exposure to solvents has caused a typical alcohol-related disease panorama.

An excess in intrahepatic bile duct cancer has been reported among rubber workers exposed to solvents by Mancuso et al.[10] An excess in mortality from lymphatic leukemias has been reported among solvent exposed rubber workers by McMichael et al.[11] Another Swedish group of workers likely to have been exposed to solvents in the past is shoe-makers. They have a cancer morbidity pattern very similar to the painters (Table 8). A display of a number of occupations from the data-set 2 with an elevated SMR for cancer of the esophagus is shown in Table 9. It is not likely that they all share an excessive intake of alcohol compared to other occupational groups.

The moderate excess of cancer in the larynx and the lungs—and probably even in the pleura—and of non malignant respiratory diseases like bronchitis, emphysema and asthma among the painters might be related to previous occupational exposure but the etiology is not known to us. It has, however, been observed that painters have a rate of positive findings on routine X-ray examinations of the lung in our surveillance programs that is higher than we would expect when comparing with people from occupations with higher dust exposure. To what extent the combined exposure to solvents and dust has a multiplicative effects is not known to us.

REFERENCES

1. Milham, Samuel, Occupational Mortality in Washington State 1950-71, US Department of HEW-CDC-NIOSH, Apr 1976.
2. Occupational mortality, The Registrar General's dicennial supplement for England and Wales 1970-72 Series DS No 1, London: Her Majesty's Stationery Office 1978.

3. Englund, A., Engholm, G., and Östund, E. Asbestos cancer in the construction industry. In Advances in Medical Oncology, Research and Education, vol. 3, Epidemiology, ed. J.M. Birch, pp 89-95, 1979. London: Pergamon.

4. Dödsorsaker 1974, Sveriges officiella statistik, Statistiska centralbyrån, Stockholm 1976.

5. National Board of Health and Welfare. Cancer Incidence in Sweden 1973, pulblished annual report.

6. Einhorn, J., Rapaport, E., Wennström, G., and Wiklund, K. Cancermiljö-registret tillgängligt för forskning. Läkartidningen 75:3415-3417, 1978.

7. National Board of Health and Welfare. The Swedish Cancer Environment Registry 1961-73.

8. Report from the National Board of Health and Welfare (Socialstyrelsen) to Work Environment Fund (Arbetarskyddsfonden) 1980.

9. Axelson, Olav et al. A case referent study on neuropsychiatric disorders among workers exposed to solvents. Scandinavian J of Work, Environment & Health, 2, pp 14-20, 1976.

10. Mancuso, T.F. and Brennan, M.J. 1970. Epidemiological considerations of cancer of the gall bladder, bile ducts and salivary glands in the rubber industry. J Occup Med. 12:333-341.

11. McMichael, A.J., Spirtas, R., Kupper, L.L., and Gamble, J.F. Solvent exposure and leukemia among rubber workers: An epidemiologic study. J Occup Med. 17:234-239, 1975.

INDEX

A

abietic acid 16, 26
acetone 44, 94, 111, 127
ACGIH 59
achanthoma 98, 100, 104, 110
  achanthomatosis 102
acrylates 28
acrylics 16, 123, 125, 127
adenomatous hyperplasia 86
adipic acid 16
adipose tissue 146
albumin 161-163, 169
alcoholism 178
alcohols 46, 132, 133
aliphatics 53, 57, 58, 123, 132, 133
alkyd 16, 17, 27, 70, 93, 123
aluminum
  powder 4
  silicate 3, 4
amino
  resins 27, 123
  -urea 17
amorphous form 9
anesthetics 157
antimony oxide 5
aromatics 46
  azo dyes 14, 15
  hydrocarbons 53, 57, 124, 132,
    133, 157
asbestos 3, 111, 120, 127
asthma 177

B

barium sulfate 3, 5
baryles 123
benzene 97, 121, 124
benzoyl peroxide 15, 29, 31
beta-2-microglobulin 161, 163, 169
bichromates 86

biocides 30, 37
bis-chloromethyl ether 73
butyl glycidyl ether 20

C

cadmium 2
  sulphide 83
  yellow 91
calcium carbonate 3, 5, 123
carbon
  black 5, 14
  tetrachloride 157
carcinogenesis 4, 110
carcinomas 101, 110
  squamocellular 98, 101, 106-109
cellulosics 16, 27, 123
central nervous system 116
cerebrovascular disease 177
china clay 4
chlorinated rubber 16, 18, 28
chloroform 157
chromates 86, 156
chromium 2, 86, 111, 127
  orange 91
  yellow 91
chronic,
  bronchitis 155, 177
  lymphatic leukemias 182
cirrhosis,
  liver 176, 178
clay 123
cobalt naphthenate 31
copper,
  oxide 6
  salts 30
creatinine 161
cristobalite 3
cyclohexanone 44, 66
  cyclohexanol 49

D

dermatoses 116
diacetone 44
   alcohol 49, 65
diatomaceous earth 3, 9
dibutyl phthalate 33
diethyl,
   ethanolamine 33
   phthalate 33
diethylene glycol monethyl ether 49
dimethyl acetone 66
Di-N-butyl sebecate 35
dioctyl phthalate 34
dipropylene glycol monomethyl ether 49
displasia 86, 100
   squamous 100, 102-104
drying oils 15

E

emphysema 177
epichlorohydrin 16, 20
Epon® 19
epoxy 16, 19, 73, 97, 125, 127
   resins 16, 29
erythrocytes 164-166
esophagus 177, 181-183
estergum 16
esters 46, 132, 133
ethanol 44, 94
ethers 132, 133
ethyl,
   acetate 44, 123
   alcohol 65, 123
   benzene 64
ethylene 15
   combinations of 44, 49, 67
excretion,
   cell 167, 169
   protein 167, 169
exposure,
   solvents 160

F

fibrogenic 3
fibrosarcomas 78, 80-84, 90
   monomorphic 79
formaldehyde 30, 35, 73, 155
fungicide 37

G

glomerular disease 158, 159
   filtration 161, 168, 169
glue sniffing 158
glycols 46

H

haematite 6
hazardous ingredients
   fillers, inorganic particles 135, 138
   metals, metallic compounds 133-138
   resins 136, 138
   solvents 132-138
health effects,
   solvents 63-67
2-heptanone 66
hexamethylene diisocyanate 24
hexylene glycol 49
hyperplasia 100

I

intrahepatic bile ducts 181, 182, 184
iron oxide 6, 14, 84, 91, 123
ischemic heart disease 177
isobutane 127
isobutanol 44, 111
isobutyl 67
   acetate 44
   alcohol 65
isophorone 66
isopropanol 44, 65
isopropyl
   acetate 44
   alcohol 123

K

kaolin 4
kerosene 63
ketones 46, 133
kidneys 158, 169

L

lactol spirits 59
larynx 181
latex 71
lead 2, 3, 14, 111, 123, 127
   chromate 7, 80
   naphthenate 32
   tetraoxide 7, 14
leucocytes 165, 166
leukemogenic 124
linseed oil 26
liver 157, 176, 178
lung 156, 181

M

magnetite 7
maleic anhydride 16
malignant diseases 180
manganese naphthenate 32
mercury 111, 127
metals 133
   metallic compounds 133
methanol 44
methyl,
   alcohol 65
   cellulose 35
   ethyl ketoxime 29, 36
   ethyl ketone 44, 66, 123
   isobutyl ketone 44, 66
   methacrylate 15, 22, 28
methylene chloride 46, 47, 142, 147, 148
mica 2, 3, 8
microscopy 161
mineral spirits 46, 59, 63
molybdenum orange 82, 91
mortality statistics 175-180

N

naphthas 59, 63
n-butanol 44
n-butyl,
   acetate 44, 67
   alcohol 65, 123
   ketone 111, 127
neoplasia 156
nephrotoxicity 157, 158
neurosis 178
n-hexane 63, 111, 127
NIOSH 59
nitrocellulose 20, 94
non-malignant 178
Novolac® 21

O

olive oil 98
OSHA 59
osmolality 161

P

panniculus carnosus 79
papillomas 98, 105, 110
para-nitroaniline 13
Parlon 18
parquet lacquer 73
pentachlorophenol 73
pentaerythitol 16
phenol mercuric acetate 36
phenolics 16, 21, 27
phthalic anhydride 16
phthalocyanine blue 11
pigments 2-11
   anti-fouling 6
   fire-retardant 5
   inorganic 4, 77
   organic 11
   red 12, 13
   violet 12
pneumoconiosis 3, 8
polyacrylates 71

polyamides 16
polymerization 15, 31
polyester 16
polyisocyanates 28
polymethyl methacrylate 22
polymethylphenyl siloxane 25
polyurethanes 23, 125, 127
propylene 15, 49
proteins 161
psychosis 178

Q

quartz 3

R

renal concentrating ability 168
resins 15, 16, 125
respiratory irritant 124
rhabdomyosarcomas 78, 80-84, 87-89
rosin 16

S

sebacic acid 16
siderosis 14
silicas 3, 123
silicone 16, 25, 28
silicosis 3, 8
SMR values 174
solvents,
    alcohol 55, 56, 65
    aliphatic 55-59, 63
    aromatic 55-58, 64
    ester 55, 56, 61, 67
    ether 55, 56, 61, 67
    health effects 63-67
    ketone 55, 56, 60, 66
    TWA 59

styrene 15, 111, 121, 145, 149, 150
    -butadiene 16, 25, 28
    expoxide 97
    oxide 97, 100, 101, 110
suicide 179

T

talc 2, 3, 9, 111, 123, 127
tert-butyl alcohol 49
tetrachlorophenol 30, 37
tetra-fluoroethylene 15
titanium,
    dioxide 2, 3, 9, 123, 127
    oxide 85, 91
toluene 44, 58, 64, 111, 123
    diisocyanate 24
toxicity 3, 116
tributyl tin chloride 30, 37
trichloroethylene 143
tridymite 3
triethylene tetramine 156
trimellitic acid 156
trimethylolpropane 16

U

uptake 141, 146
urea formaldehyde 94
urethane 16, 123
urinalysis 161

V

varnishes 94
Versamid® 22
vinyl, 16, 27
    acetate 25
    chloride 15

W

white spirit 70

X

xylene 44, 58, 64, 111, 123, 144, 151

Z

zinc, 2
  chromate 10, 14, 81
  dust 10
  oxide 10, 14
  yellow 91
zymbal gland 99

About the Series Editor

Myron A. Mehlman, Ph. D.

Dr. Myron A. Mehlman is the Director of Toxicology and Manager of the Environmental and Health Sciences Laboratory at Mobil Oil Corporation. Born in 1934, Dr. Mehlman was educated at the City College of New York (B.S. 1957), the Massachusetts Institute of Technology (Ph.D. 1964), the University of Wisconsin (Post-Doctoral Fellow, Institute for Enzyme Research, 1967), and Harvard Business School (Program for Health Systems Management, 1974). His academic appointments include Associate Professor of Biochemistry (1967-1969) at Rutgers University and Professor Biochemistry (1969-1974) at the University of Nebraska. He was appointed Adjunct Professor of Medicine at the Mt. Sinai School of Medicine in 1980.

Dr. Mehlman was Chief of Biochemical Toxicology (1972-1973) at FDA, Special Assistant for Toxicology, Environmental Affairs, and Nutrition (1973-1975) at Office of Assistant Secretary for Health, HEW, and Special Assistant for Program Planning and Evaluation, and Interagency Liaison Officer, Office of Director at National Institutes of Health (1975-1977).

In addition to serving as Chairman of the First and Second National Meetings of the American College of Toxicology, he has chaired symposia at Rutgers University, the University of Nebraska, FASEB, NIH, and the FDA. His professional memberships include the American Society of Biological Chemists, the American Physiological Society, American Institute of Nutrition, American Society for Experimental Therapeutics and Pharmacology, the American Chemical Society, the Society of Toxicology, the American College of Toxicology and the American Industrial Hygienist Society.

Dr. Mehlman is also the founding editor of the Journal of Toxicology and Environmental Health and the Journal of Environmental Pathology and Toxicology, and has been a series editor for Advances in Modern Nutrition, Advances in Modern Toxicology, Symposium on Metabolic Regulation, and Advances in Modern Environmental Toxicology. From 1977-1979 Dr. Mehlman was the first and founding president of the American College of Toxicology.

Since 1962, Dr. Mehlman has published 162 articles in the fields of biochemistry, toxicology, nutrition and human health. He has edited and co-edited approximately 18 books.

SECTION IV
REPRINTS

Reprinted by permission.
Scand. j. work environ. & health 4 (1978) 19—45

Long-term exposure to jet fuel

II. A cross-sectional epidemiologic investigation on occupationally exposed industrial workers with special reference to the nervous system

by BENGT KNAVE, M.D.,[1] BIRGITTA ANSHELM OLSON, M.Sc.,[1] STIG ELOFSSON, Ph.D.,[3] FRANCESCO GAMBERALE, Ph.D.,[1] ANDERS ISAKSSON, D.Sc.,[2] PER MINDUS, M.D.,[2] HANS E. PERSSON, M.D.,[2] GÖRAN STRUWE, M.D.,[2] ARNE WENNBERG, M.D.,[2] and PETER WESTERHOLM, M.D.

KNAVE, B., ANSHELM OLSON, B., ELOFSSON, S., GAMBERALE, F., ISAKSSON, A., MINDUS, P., PERSSON, H. E., STRUWE, G., WENNBERG, A. and WESTERHOLM, P. Long-term exposure to jet fuel: II. A cross-sectional epidemiologic investigation on occupationally exposed industrial workers with special reference to the nervous system. *Scand. j. work. environ. & health* 4 (1978) 19—45. Thirty jet fuel exposed workers selected according to exposure criteria and thirty nonexposed controls from a jet motor factory were examined, with special reference to the nervous system, by occupational hygiene physicians, psychiatrists, psychologists, and neurophysiologists. The controls and the exposed subjects were matched with respect to age, employment duration, and education. Among the exposed subjects the mean exposure duration was 17 years, and 300 mg/m³ was calculated as a rough time-weighted average exposure level. The investigation revealed significant differences between the exposed and nonexposed groups for (a) incidence and prevalence of psychiatric symptoms, (b) psychological tests with the load on attention and sensorimotor speed and (c) electroencephalograms. When the control group was selected, it was ensured that the two groups were essentially equivalent except for exposure to jet fuel. It is concluded, therefore, that the differences found between the groups are probably related to exposure to jet fuel.

Key words: epidemiology, jet fuel, occupational medicine, occupational neurology, occupational neurophysiology, occupational psychiatry, occupational psychology, organic solvents.

It is well known that exposure to petroleum distilled fuels may have effects on the nervous system in man. Numerous reports have been published on acute and chronic occupational intoxication due to

---

[1] National Board of Occupational Safety and Health, Stockholm, Sweden.

[2] Departments of Clinical Neurophysiology and Psychiatry, Karolinska Hospital, Stockholm, Sweden.

[3] Department of Statistics, University of Stockholm, Sweden.

Reprint requests to: Associate Prof. Bengt Knave, Arbetarskyddsstyrelsen, Arbetsmedicinska Avdelningen, Sektionen för fysikalisk yrkeshygien, 100 26 Stockholm, Sweden.

such fuels. Acute intoxication is followed by dizziness, drowsiness, nausea, vomiting, etc., i.e., a narcotic effect (19, 28, 39, 41). As a matter of fact, gasoline was tried, although less successfully, for general anesthesia more than 100 years ago (16). Symptoms have also been described in a case of inflight intoxication due to jet fuel fumes (13). Neurasthenia (24, 38, 44, 45, 46, 47) and polyneuropathy (9, 12, 15, 34, 38, 46) have been described as early manifestations of disease following chronic intoxication.

Recently 29 Swedish aircraft factory workers occupationally exposed to jet fuel

198

were examined (32, 33). More than two-thirds of them reported recurrent acute symptoms (dizziness, respiratory tract symptoms, heart palpitations, a feeling of thoracic oppression, nausea, headache) upon exposure at work to jet fuel vapors. Furthermore, symptoms of neurasthenia and symptoms and signs of polyneuropathy were more frequent in the exposed group than in the reference groups. These findings were interpreted by the authors as a possible effect of long-term exposure to jet fuel on the nervous system. The authors stated, however, that further studies on other groups of workers exposed to jet fuel vapors and on matched control groups of nonexposed workers were needed before the validity of this interpretation could be established. The aim of the present study was to provide additional information on such groups.

A jet motor factory was chosen for this purpose. In some areas of the factory parts of the engine are manufactured, and the workers concerned are not exposed to jet fuel. In another part of the plant motor parts are assembled, and the motor is run under simulated inflight conditions. Workers involved in these procedures are exposed to jet fuel. Thus, it was possible to compare exposed and nonexposed groups of subjects employed in the same factory.

THE WORKPLACE

Exposure to jet fuel vapors regularly occurs, particularly in the units or sites of the factory where the fuel systems of engines are produced. These fuel systems consist of a large number of single components. During the production process the components are gradually assembled to constitute the end product. During this process the various components and the fuel system as such undergo a variety of standardized testing procedures for the assessment of function, technical quality, and security. During the tests jet fuel flows through the test object in order to create a situation which can be regarded as equivalent to real flying conditions. In performing these tests, some categories of personnel are exposed to jet fuel vapors.

WORK CONDITIONS

Testing of fuel systems (component testing)

Fuel systems are tested on special premises (i.e., in test houses). The various components to be tested are attached to a testing device ("rig"). The actual test is performed and supervised by one man and consists of readings and manipulations and adjustments on the test object during the course of the test. There is a large variety of components, and the testing periods can vary from 10—15 min to a whole day, depending on the type of object and the length of the standardized testing protocol. Common test objects are: main fuel controls, after-burn pumps, exhaust nozzal controls, after-burn fuel controls, ignitian valves, pressure radio bleed controls or valves, computers, etc.

Performing these tests and control procedures is manual work requiring high standards of precision; the work implies that the person performing it is positioned with his face 20—40 cm from the test object. The object is constantly perfused with jet fuel. For a tester roughly 50 % of the total work time is spent in this manual work with direct exposure to fuel vapors. The rest of the time is spent supervising the test at a distance (from behind a window), writing protocols, etc., with a minimum of exposure. The testing personnel consistently give the same description of their work situation and practices. They circulate freely from one part of the premises to another, and they do not specialize in any particular kind of test item.

Engine testing

The engines, when assembled, are also subject to testing and control procedures under conditions simulating real flying. The engines, with their installed fuel supply systems, are put up in a stabilizing frame and run through a standardized testing procedure. To a large extent the testing is maneuvered from a protected room with distant-control systems. During some moments (connection and disconnection of fuel supply, control and cleaning of fuel filters, adjustments, etc.), the testing personnel handle the fuel system and are directly exposed to fuel vapors. The personnel consistently report that the exposure periods take roughly 35 % of their total work time. The engine tests can last from one or two to ten days. Usually they take three days.

Other kinds of work

After having undergone the testing procedure, which may be repeated after adjustments, the test object is passed on to special personnel (mechanics) for disassembling, cleaning and reassembling. These tasks require precision and are performed with the breathing zone of the worker in the immediate vicinity of the engine and component parts drenched with fuel. The personnel give the round figure of 35 % as the amount of work time spent at these tasks.

The testing of component parts, as already described, implies the use of instruments and devices for observing and registering the course of the testing procedure for evaluation purposes. The instruments need almost constant supervision and calibration for accuracy and reliability. Instrument checks are performed by special personnel who, from a practical point of view, work under similar conditions and routines as the component testers.

There is also a constant need of instrument and testing service repairs. This work is performed by one mechanic who works in the testing area. The nature of this work is very similar to that of the component testers with the difference that, when repairs are done in the testing area, the flow of fuel is regularly stopped.

EXPOSURE

Jet fuel is an aviation turbine fuel with the approximate distillation range of 50 to 250°C. Since 1972 fuel with the following composition has been used:

Aromatic hydrocarbons	12 vol. %
Olephin hydrocarbons	0.5 vol. %
Saturated hydrocarbons (paraffins, cycloparaffins, etc.)	87.5 vol. %

The aromatic hydrocarbons include benzene (0.3 vol. % of the total), toluene (1 %). The fuel contains no lead compounds.

In the present study the assessment of exposure to jet fuel vapors is based on (a) time of employment in work implying exposure, (b) analysis of work practices, and (c) measurements of fuel contents in the air of the worker's breathing zone during exposure. Time of employment in work implying exposure to fuel vapors was obtained from a personal interview and verified with the aid of the records of the company. At the interview a detailed account was obtained of the work habits and practices of the individual throughout the years during which exposure to fuel had occurred. Also, at this time, the moments of work when such exposure reasonably could occur were determined. On the basis of these interviews various moments of the work procedures were selected for quantitative measurement and analysis. The interview, moreover, constituted the basis for a gross overall approximation of the percentage of time occupied by work during which any considerable extent of exposure occurred to jet fuel vapors.

During 1975 a program of technical improvements was performed at the factory. The fuel vapor concentrations were thus decreased by improved ventilation and other protective arrangements, particularly in the areas of component testing. The results of these improvements are illustrated in fig. 1.

Quantitative measurements

In assessing exposure, we used measurements made of jet fuel concentrations in the air before the technical improvements were accomplished. Measurements were also performed during work procedures which had not undergone technical change.

Air concentrations of fuel vapors were measured with a flame ionization detector (Ratfisch Instrument, Model RS 5). These measurements were performed to give a result relevant to exposure during work. The test person had an air sampler attached to him during work, and the sampling point was positioned immediately in front of his nose. The teflon sampling

200

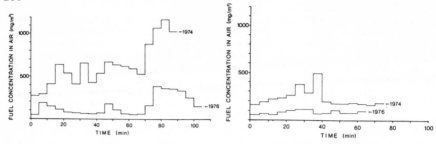

Fig. 1. Levels of fuel concentration in the worker's breathing zone during component testing with adjustment of a high pressure body (figure on the left) and a computer (figure on the right) as the test object in 1974 (upper lines) and 1976 (lower lines). The levels are given as the time-weighted average for each 5-min period.

tube measured 4 mm in internal diameter and 7 m in length. It was considered not to interfere with the routine course of the work procedure. The air flow in the tube during sampling was 2 l/min.

In November 1974 measurements of air concentrations of jet fuel were performed during some component-testing procedures. The test objects were selected to give a representative mixture of the work situation for exposed personnel. The results as to component testing, engine testing, and relevant work procedures performed by mechanics are shown in table 1.

In the table the calculation of the average of the whole measurement is based on the time-integrated average concentration for each 5-min period of current measurement. For the registration of the time-integrated averages, an integrator device (Rikadenki Computer Recorder) was used.

Table 1. Jet fuel concentrations (mg/m³) in the breathing zones of personnel at work.

Test object	Time (t) (min)	Maximum	Minimum	Mean (x̄)
During component testing				
Adjustment of flow control	40	343	157	209
Adjustment of computer	77	488	160	229
Adjustment of high pressure body	55	128	51	85
Adjustment of exhaust nozzal control	90	1,168	270	630
Adjustment of after-burn pump vent	20	594	422	510
Flow testing spread-ring	19	221	178	200
End control after-burn vent	19	824	284	580
Adjustment of main fuel control	97	3,214	111	925
Control of component in after-burn pump	87	413	41	125
During engine testing				
Final control of fuel filter	33	237	58	140
Regular control of fuel filter	54	272	29	110
Disconnecting fuel tubing, emptying of engine	30	240	41	147
During routines implying exposure to fuel vapors				
Collection of spill fuel	10	— a	— a	480
Repair of leaking fuel tube	10	— a	— a	149
Trimming of locking device	40	253	142	185
Change of filter	55	3,226	56	974

a Not given.

Assessment of exposure

The exposure of different categories of personnel can be expressed as a time-weighted average air concentration in the breathing zone during the time periods when exposure occurs. This average value (C) was calculated to be as follows: 423 mg/m³ for the component testers, 128 mg/m³ for the engine testers, and 185 —248 mg/m³ for the mechanics. The figures for the mechanics are given as the range due to the fact that the group was heterogeneous with respect to exposure conditions. C is calculated from the results of the performed measurements of air concentrations (table 1), and the registered time length of each test operation, with the following formula:

$$C = \frac{\Sigma \ (t x \bar{x}) \ n}{\Sigma \ t} \quad ^4$$

The entity (C) bears the character of a crude approximation and is demonstrated with the purpose of describing the levels of exposure in terms of magnitude.

EXAMINED GROUPS

Exposed workers

A committee was formed with representatives of the management, the trade unions, and the health department of the plant. The committee, acting on our instructions, selected 30 of the most heavily exposed workers from all the employees in regular or intermittent contact with jet fuels. The selection was based on accurate knowledge of the work conditions of all the jobs concerned. The exposure criteria were intensity and duration of exposure.

There was a consensus of opinion within the committee on the selection, and all

selected individuals agreed to participate in the study. In this group 15 were fuel system testers (5 foremen), 7 engine testers (2 foremen), and 8 mechanics (4 foremen). The fuel system testers were considered to be more exposed than the engine testers and mechanics. The duration of exposure varied between 2 and 32 years (mean = 17.1, median = 18.5).

Control groups

Originally, an age-matched control group of 30 subjects (age range 27—66 years, mean = 46.4, median = 44.0) was selected by the former plant physician. In this selection procedure "general state of health" was also used as a matching criterion. The level of education was disregarded. Since this procedure is open to objections, another control group had to be selected.

The final control group was chosen under the supervision of the Department of Statistics, the University of Stockholm. Subjects eligible for inclusion in the control group were drawn from the employment records according to the following criteria: (a) no exposure to jet fuel, (b) same age (± 1 year), and (c) similar duration of employment (± 1 year for short and ± 5 years for long durations) as each of the exposed subjects. From this pool of subjects (n = 162), one control candidate for each of the exposed foremen and three control candidates for each exposed worker were drawn by lots. The level of education and any trade union activity were established in this subgroup, since these factors were considered important to match. All 30 control subjects thus selected participated in the study.

The exposed group and the final control group, pairwise matched, were each composed of 19 workers and 11 foremen, all males, with an age range of 27—66 years. The mean age was 46.4 in the exposed group and 46.2 in the control group. The median age was 43.5 years in both groups. The exposed group had been employed 17.7 (mean) years and the control group 19.8 (mean) years. The corresponding median values were 19.0 and 21.0 years, respectively.

4 The formula is applied to the measurement results of component testing in table 1 as follows: [(40 x 209)/504] + [(77 x 229)/504] + [(55 x 85)/504] + [90 x 630)/504] + [(20 x 510)/504] + [(19 x 200)/504] + [(19 x 580)/504] + [(97 x 925)/504] + [(87 x 125)/504 = 422.6 mg/m³.

202

METHODS

Medical history, standardized interview and neurological examination (investigator: B. K.)

Information was collected on heredity, previous and current health, occupational history, tobacco smoking, use of alcohol, etc., as well as on acute symptoms associated with exposure to jet fuel. Data on the incidence of symptoms such as neurasthenia, anxiety, and/or mental depression and on confounding and effect modifying factors were collected with a standardized interview, the contents of which were agreed upon in a Scandinavian meeting on health hazards in the use of solvents (3). Moreover, in the present investigation we were able to obtain further information from the medical records of the factory. Nearly all the subjects had been employed at the plant almost all their adult life (cf: mean age about 46 years and mean duration of employment about 19 years), and during this time they had been referred to the plant physician (who had been employed at the factory for 27 years) when ill. Therefore symptoms of neurasthenia, anxiety and/or mental depression were recorded and evaluated also from notations in the medical records of the factory health department.

When polyneuropathy was being studied, we used the standardized questionnaire applied in the first jet fuel investigation (32, 33) and supplemented the clinical neurological examination with an examination especially designed to detect early signs indicative of polyneuropathy. A basis for such an evaluation was developed by Lindblom in studies on uremic patients with subclinical and manifest polyneuropathy (7) and was later used in investigations on industrial workers (29, 30, 31, 32, 33, 40). Findings are scored between 0 = normal, 1 = mild changes, 2 and 3 = different severity of manifest disease. A score of 1 may be "normal," however, particularly for the elderly.

Psychiatric interviews and ratings (investigators: P.M. and G.S.)

The interview and first evaluation. All interviews were made during the summer 1976 by a psychiatrist who had no knowledge of the examined workers' previous diseases or symptoms. The workers were interviewed in random order. The interview lasted about 45 min. The medical history was first obtained; in it life events and illnesses promoting mental symptoms were emphasized. The contents (items) of the interview are shown in part A of the appendix. The second part of the interview concerned the prevalence of certain psychiatric symptoms. A record of 37 different symptoms (items) was obtained with a modified version of the Comprehensive Psychopathological Rating Scale (CPRS) (1). Each item was rated on a 7-grade scale of increasing severity; the rating depended on the intensity, frequency and duration of the symptoms in question. The items are shown in part B and the rating scale in part C of the appendix.

Taping procedures and second evaluation. The interviews were recorded on tape. The psychic symptoms were reevaluated by the interviewer on the basis of the tape recording, and then separately by another psychiatrist with no previous contact with the examined workers. The scores of each item (individual score) were added and the sum formed the measure of the prevalent symptoms in each subject. The mean of the two independent secondary evaluations was regarded as the measurement result.

Assessment of interrater reliability in the evaluation produced the following results: (a) The product-moment correlation between the scores of mental symptoms rated during the interview and those after evaluation of the tapes by the interviewer was $r = 0.79$. (b) The product-moment correlation between the interviewer's evaluation of the tapes and a similar rating by the other psychiatrist was $r = 0.87$ (assessed for 16 of the 60 tape recordings). (c) When observer bias was estimated, information on work conditions in the last 42 tapes was withheld from the second psychiatrist. In this situation the correlation between the ratings made by the interviewer and the second psychiatrist was r = 0.73. The examinee's work conditions were misjudged by the second psychiatrist in half of the cases. After evaluation of the tapes the overall credibility of the reports given during the interview was found to be fairly good and not related to the individual's score.

The first part of the interview was summarized in a medical record, on the basis of which five items (part A of appendix) were rated (part C of appendix) independently by both psychiatrists. The resulting score was considered to be an index of the individual's liability for mental symptoms. For the comparison between the exposed and control groups, factors concerning previous mental health (item 3) were omitted. The differences in the index between matched pairs were calculated. In order to reduce random error, we considered only differences larger than one (2). With this evaluation procedure, the accordance between raters was good. (In 25 of 30 cases there were similar conclusions.)

Statistical procedure. The difference between the scores of the matched pairs was tested by the sign test. In addition, the difference in mental symptoms between the groups was tested by means of ranking the individuals (Mann—Whitney). The quantitative difference in psychic symptoms between matched individuals was tested by Student's t. Similar tests for the difference between the groups were also used for various categories of items.

Psychological tests (investigators: F. G. and B.A.O.)

Six performance tests were carried out individually by each subject within 1 h. The subjects were studied in such an order as to obtain comparability between the exposed and nonexposed groups with regard to time of day and day of week for the examination. The tests were performed by each subject in the order in which they are presented.

RT Addition (RT = reaction time). The RT Addition task was carried out with the aid of an electronic instrument and a display-control panel. The test was conducted so that three single-digit numbers were presented simultaneously to the subject for 1 s. The subject was to add the three numbers and indicate the result on a keyboard. Forty-six trials were performed. The mean reaction time and standard deviation were calculated for each subject based on the last 40 trials.

Simple RT. The Simple RT task was performed with a stimulus/response panel placed horizontally to the subjects. The subjects were instructed to respond to a visual signal (stimulus) by pressing a switch with their fingertips as quickly as possible. The interval between stimuli averaged 3.75 s with a maximum variation of ± 1.25 s. Each session lasted about 10 min, during which a total of 160 stimuli were administered; thus there were 16 signals/min. The signals were administered electronically, and an electronic timer was used to record the reaction time. The mean reaction time and standard deviation were calculated for each subject based on the last 9 min of the task. Furthermore the mean reaction time was calculated for each time block (1-min period) to be used in an analysis of change in performance over time.

Memory test (recognition). For the testing of memory recognition, the same apparatus was used as for RT Addition. A series of 20 meaningless but pronounceable three-letter syllables, each lasting 2 s and at an interval of 1.5 s, was presented to the subject on the display panel twice. The same 20 syllables were thereafter presented to the subject once again randomly mixed with another 20 syllables which had not appeared before. The task of the subject consisted of indicating whether each of the 40 syllables had or had not been presented previously. The number of correct answers was recorded.

Manual dexterity (Santa Ana). The manual dexterity test consisted of a board (22 × 75 cm) with four rows, each row consisting of 12 square holes. Into each hole a knob could be fitted, the upper part of which was cylindrical. The task consisted of lifting the knobs one at a time from the hole, turning them 180°, and putting them back in the hole. Five trials, each lasting 30 s, were carried out, two with each hand and one with both hands simultaneously. The number of knobs correctly replaced was recorded.

Perceptual speed (Bourdon—Wiersma). A modified version of the Bourdon—Wiersma Vigilance Test was used to test perceptual speed. The test material consisted of one full page with groups of 3, 4 or 5 dots arrayed in 30 lines, with 25 dot groups per line. The task of the subjects was to draw a line over each group of four dots, working from left to right, line by line. The subjects were instructed to work both rapidly and accurately. The task lasted 5 min. The mean time per line was calculated. Furthermore mean performance was calculated for each time block (1-min period) to be used in an analysis of change in performance over time.

Memory test (reproduction). When memory reproduction was tested, the same apparatus as for RT Addition and memory recognition was used. A series of letters and numbers was presented on the display panel during 4 s. The task of the subject was to reproduce the combination as accurately as possible. A total of 17 trials with combinations of increasing length were performed successively. The number of correctly reproduced combinations and the number of correctly reproduced elements were recorded.

The Simple RT, manual dexterity, and perceptual speed tests have all been used in previous studies (17, 18, 22, 36). The two memory tests and RT Addition were specially devised for the present study.

Neurophysiological examinations

(a) Electroencephalograms (investigators: A.W. and A.I.)

The electroencephalograms (EEGs) were recorded by means of an 8-channel Siemens—Elema EEG machine. The subjects were awake in a semireclined position with eyes closed during the recording. The electrodes were placed according to the 10/20 system. For computer analysis six derivations were stored on a 6-channel

analogue FM tape recorder (Precision Instrument): F7-T3, T3-C3, P3-01 and F8-T4, T4-C4 and P4-02.

The paper recordings were evaluated in the following way: (a) They were visually examined according to the clinical routine procedure. (b) From each EEG recording a typical 10-s section (8 channels) was selected. These sections were ranked without any knowledge of which subjects they belonged to. This ranking procedure was carried out so that the EEGs showing a distinct, stable alpha activity were given low ranks while higher ranks were

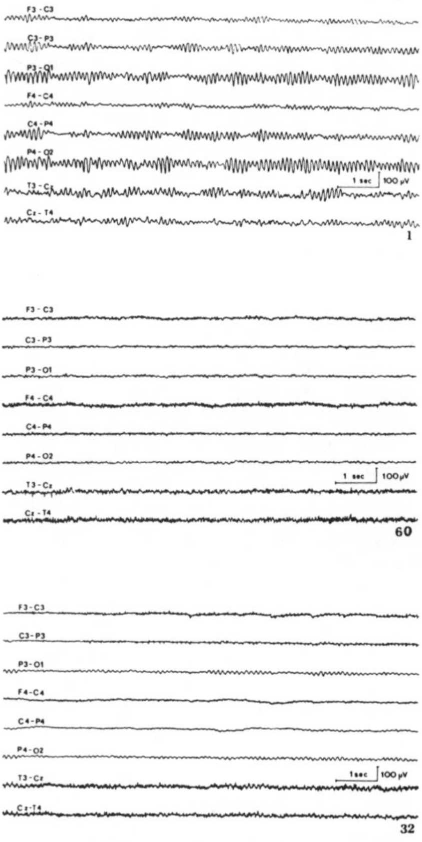

Fig. 2. EEG recordings from eight leads with ranks 1, 32, and 60 assigned after visual inspection.

assigned to EEGs with a decreasing amount of rhythmic activity. Fig. 2 shows three examples with different ranks (1, 32 and 60, respectively).

The recorded signals were computer processed by a method called Spectral Parameter Analysis (SPA) (48, 49). The SPA is a linear filter model adapted to the EEG signal, and it describes the spectral properties of the EEG by a limited number of components. As a rule, each component corresponds to a certain type of activity in the EEG. For this reason the components are labeled delta, alpha and beta, as in usual clinical EEG nomenclature. Each component is characterized by three parameters: (a) the peak frequency (f, Hz), (b) the bandwidth (σ,Hz), and (c) the power (G, percentage of total power). By definition the peak frequency of the delta component is zero; therefore this component is described by two parameters only. The total power spectrum of the EEG is thus characterized by eight parameters (2 + 3 + 3), as can be seen in fig 3. To these another parameter is added which expresses the ratio between the mean square deviation of the model to signal adaptation and the variability of the spectral properties of the signal (a model fit parameter). For good adaptation this parameter should be smaller than 1.00, and it decreases with increasing degree of adaptation. The main features of the SPA procedure have been described by Isaksson (26).

For the purpose of the present study, artifact-free EEG signals for computer analysis were provided by the visual selection of EEG sections of 10-s duration and free of artifacts such as muscle activity, arterial pulsations, eye and electrode movements. After the SPA analysis of these sections the parameters in question were inspected. Some of the results were finally canceled due to the following criteria: (a) poor model adaptation, i.e., a model-fit larger than 1.25, (b) strongly deviating parameters, i.e., negative power parameters, double peaks, etc.

After these selection procedures 8—14 sets of parameters remained for each subject from the leads P3-01 and P4-02 for 22 out of the 30 couples. This material was then subject to the following statistical procedures.

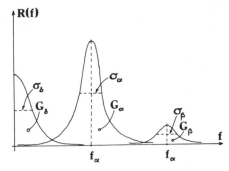

Fig. 3. Spectral power R(f) as a function of frequency (f).

1. For each parameter the mean and standard deviation were calculated for each subject. Then the sets of parameter means were used to characterize the subjects. Based on each parameter separately, the hypothesis "equal means for exposed and control subjects" was tested with the analysis of variance, and the differences in standard deviations were tested with Fisher's F test.

2. The parameter means of all the subjects were ranked for each parameter separately. In addition a total ranking was performed in which each subject was ranked according to the sum of all his parameter ranks. The differences in rank between the exposed subjects and the controls were tested by means of the Mann—Whitney rank test.

3. The original sets of parameters were used to characterize the subjects. For each matched couple, it was determined whether the exposed subject was significantly different from the control. This procedure was carried out for each parameter separately by means of the analysis of variance. For statistical references the reader should consult Lindgren's *Statistical Theory* (35).

(b) Peripheral nervous system examinations (investigator: H.E.P.)

The functional state of peripheral nerves in the examined subjects was assessed by means of measuring conduction velocities and compound nerve action potential amplitudes in sensory nerves, conduction velocities in motor nerves, and sensation thresholds of vibration in the extremities.

Conduction velocities and nerve action potential amplitudes in peripheral nerves. Four peripheral nerves (n medianus, n ulnaris, n peroneus and n suralis) were investigated with electroneurographic (ENeG) methods (fig. 4) recommended in a Scandinavian meeting on health hazards in the use of solvents (3). The ENeG methods have earlier been used in studies on jet fuel exposed personnel (32, 33) and on other industrial workers (29, 30, 31, 40), but the present investigation was more extensive.

The sensory nerve conduction velocities (SCVs) and nerve action potential (NAP) amplitudes were measured in a distal (index finger-wrist; dse-S1 in fig. 4) and a proximal section (wrist-elbow; S1—S2 in fig. 4) of the left median nerve, and in a distal section of the left ulnar (dig V manus-wrist; dse-S1 in fig. 4) and sural nerves (mid lower leg — ankle region;

S2—S1 in fig. 4). Orthodromic measurements were made of all SVCs and NAPs with the exception of those of the sural nerve, in which antidromic methods were used. The measurements of the maximal conduction velocities (MCVs) in the motor nerves were made in the left median (elbow-wrist; S2—S1 in fig. 4), ulnar (elbow-wrist; S2—S1 in fig. 4) and peroneal nerves (knee-ankle; S2—S1 in fig. 4). The conduction velocity of the slow motor fibers (CVSF) was determined in the left ulnar nerve (elbow-wrist; S2—S1 in fig. 4). An anatomical illustration is presented in fig. 4.

The measurements of the MCV, SCV and NAP were made according to routine clinical methods (10, 20, 37, 42). The technique of measuring the CVSF in the ulnar nerve has previously been described by Seppäläinen and Hernberg (43). All the recordings were made with surface electrodes. Two MCV and CVSF values and four SCV and NAP values were recorded from each nerve segment of the examined persons.

Sensation thresholds of vibration in the extremities. The vibration thresholds were determined at three different locations of the extremities, i.e., the bony parts of the dorsum of the foot (represented in the tables by tarsal), lower leg (tibial) and wrist (carpal).

The measurement of vibration thresholds has earlier been used in investigations on industrial workers (29, 30, 31, 32, 33, 40). An electromagnetic biothesiometer with a plastic stimulator shaft, 6 mm in diameter, provided a 100-Hz sine wave stimulus with a variable amplitude (0—25 μm). The stimulus amplitude was displayed digitally in micrometers peak to peak with the aid of an accelerometer mounted on the shaft of the stimulator. The stimulator was applied in a vertical position and perpendicular to the skin surface with the pressure of its own weight (440 g) on the skin surface over the underlying bone where the subcutaneous tissue was thinnest. This application minimized variation in repeated tests.

Differences in the parameters of ENeG and vibration thresholds between matched subjects were tested by parametric methods.

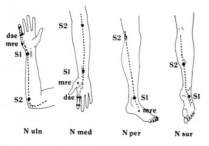

Fig. 4. Schematic illustration showing the positions of the stimulating and recording electrodes on the four investigated peripheral nerves. (N uln: nervus ulnaris; N med: nervus medianus; N. per: nervus peroneus; N sur: nervus suralis; S: position of stimulating and/or recording electrodes; dse: position of distal stimulating electrodes; mre: position of muscle respoι..e recording electrodes)

Table 2. Acute symptoms.

Symptom	Fuel system testers (n=15)	Motor engine testers and mechanics (n=15)	All subjects in the exposed group (n=30)
Dizziness	9	6	15
Fatigue	5	8	13
Headache	4	3	7
Nausea	1	3	4
Palpitations, thoracic oppression	3	1	4
Respiratory tract symptoms	2	1	3

General comment

In most of the investigations (i.e., when possible) the investigators did not know whether the person being examined belonged to the exposed or nonexposed group. The different investigations were performed and evaluated separately and independently of each other.

RESULTS

Medical history, standardized interview and neurological examination

(a) Acute symptoms

Twenty-one of the 30 exposed subjects stated that they had experienced recurrent acute symptoms upon exposure (table 2). The symptoms consisted of dizziness, headache, nausea, respiratory tract symptoms ("pain upon inhalation," "feeling of suffocation," "slight cough"), palpitations, and a feeling of thoracic oppression. Thirteen subjects also reported fatigue during work with jet fuel and afterwards, especially in the evenings. Altogether only four subjects in the exposed group did not report acute symptoms on exposure. As can be seen in table 2, no remarkable differences were found between the component testers on one hand and the motor engine testers and the mechanics on the other.

(b) Nonacute symptoms

Incidence of neuropsychiatric symptoms during employment at the factory. Nearly all of the examined subjects had been employed at the plant almost all their adult life. The incidence of neuropsychiatric ill health during this time was determined. The determination was based on the medical history and the standardized interview, but also on notes in the plant physician's medical records. Only events before September 1974 were considered in the incidence studies since at that time jet fuel was recognized as a potential health hazard in Sweden.

The results are shown in tables 3 and 4, in which comparisons are made between the exposed group and the control groups, and the differences are expressed as significance levels (p). Comparisons are made between the exposed group (n = 30) and the original control group, as well as with the final control group (n = 2 × 30). It appears from the tables that the two control groups are very similar in regard to incidence of symptoms. The incidence of symptoms in the exposed group differed significantly, however, from those of the control groups, as evaluated from the medical history and standardized interview, as well as from the plant physician's medical records.

Significant differences (based on groupwise as well as pairwise comparisons) were found when symptoms of neurasthenia, anxiety and/or mental depression were recorded whenever information was available on their existence and whenever

Table 3. Symptoms of neurasthenia, anxiety and/or mental depression in the exposed group and the control groups as recorded from medical history and standardized interview (upper row of figures) and medical records of factory health department (lower row).

Symptom	Exposed group (n=30)	Final control group (n=30)	Statistical difference (p)	Original control group (n=30)	Statistical difference (p)	Original and final control groups (n=60)	Statistical difference (p)
Neurasthenia, anxiety and/or mental depression	24 23	10 10	0.0008 0.0018	11 12	0.0017 0.0089	21 22	0.0001 0.0008
Diagnosed and treated by physician	19 23	7 10	0.004 0.0018	8 12	0.0095 0.0089	15 22	0.0009 0.0008
On several occasions	14 19	4 6	0.011 0.0017	4 4	0.011 0.0002	8 10	0.0014 0.0001
Anxiety and/or mental depression on one occasion	3 4	3 4		7 7			

Table 4. Symptoms of neurasthenia in the exposed and control groups as recorded in the standardized interview.

Symptom	Exposed group (n=30)	Final control group (n=30)	Statistical difference (p)	Original control group (n=30)	Statistical difference (p)	Original and final control groups (n=60)	Statistical difference (p)
Fatigue	13	1	0.0008	1	0.0006	2	< 0.0001
Depressed mood, lack of initiative, etc.	10	1	0.0076	1	0.0076	2	0.0003
Dizziness	10	2	0.024	1	0.0076	3	0.0010
Palpitations, thoracic oppression	9	1	0.015	2	0.046	3	0.0031
Sleep disturbances	9	2	0.046	1	0.015	3	0.0031
Headache	5	1	0.20	0	0.061	1	0.025
Memory impairment	5	0	0.061	3		3	0.15
Irritability	4	1	0.35	2		3	0.33
Respiratory tract symptoms (feelings of suffocation, etc.)	3	0	0.24	2		2	0.42
Sweating	3	2		0		2	0.42

Table 5. Symptoms of gastritis in the exposed and control groups as recorded in the standardized interview (upper row) and plant physician journals (lower row).

Exposed group (n=30)	Final control group (n=30)	Statistical difference (p)	Original control group (n=30)	Statistical difference (p)	Original and final control groups (n=60)	Statistical difference (p)
15 22	10 15		10 16		20 31	 0.08

Table 6. Eye irritation in the exposed and control groups as recorded in the standardized interview.

Exposed group (n=30)	Final control group (n=30)	Statistical difference (p)	Original control group (n=30)	Statistical difference (p)	Original and final control groups (n=60)	Statistical difference (p)
9	1	0.015	3		4	0.0081

the subjects had consulted a physician and been treated for the symptoms. Significant differences were also found when the number of subjects who had experienced ill health on several occasions were compared. No difference was found for the incidence of one single attack of anxiety and/or mental depression.

Table 4 shows the incidence of symptoms in the exposed and control groups as recorded from the standardized interview. Among the neurasthenic symptoms (including the neurovegetative ones) the most obvious differences between the exposed and nonexposed groups were found for fatigue, depressed mood, lack of initiative, and dizziness. Other symptoms which differed markedly between the groups were palpitations, thoracic oppression, sleep disturbances, and headache.

Incidence of gastritis and eye irritation in the exposed and control groups during employment at the factory. In the standardized interview, questions were included on symptoms of gastritis and eye irritation since gastritis is a psychosomatic disease and eye irritation has been described as one of the first symptoms to appear on exposure to relatively low concentrations of gasoline in experimental studies (14). The results are shown in tables 5 and 6, respectively. Gastritis symptoms were more frequent in the exposed group than in the nonexposed groups, especially when the notations in the plant physician's journals were evaluated and the two control groups were combined to form the reference group (p < 0.1). As for eye irritation, significantly more subjects in the exposed group complained of symptoms when compared

with subjects of the genuine control group (p < 0.05) and the two control groups together (p < 0.01). The eye irritation, as described by the examined subjects, could be defined as symptoms of chronic conjunctivitis and, thus, were not necessarily associated only with exposure to jet fuel.

Occurrence of symptoms and signs possibly indicative of polyneuropathy. The occurrence of symptoms possibly indicating polyneuropathy ("restless legs," muscle cramps, diffuse pain in the extremities, distal paresthesia and numbness, and paresis), was determined from data collected with the polyneuropathy questionnaire described elsewhere (32). A symptom was scored as "positive" when the subject experienced it more often than once a month. It should be emphasized that "positive symptoms," according to these criteria, do not necessarily indicate manifest disease in terms of a clinical evaluation (32). Subjects who experienced at least one of the symptoms were classified as "positive." As shown in table 7, subjects with such symptoms were more frequent in the exposed group than in the control groups (p < 0.1).

In the standardized neurological examination signs scored 0 and 1 were found. As shown in table 7, no significant differences were found between the exposed and nonexposed groups.

(c) *Possible confounding and effect modifying factors*

As to possible confounding and effect modifying factors (use of tobacco and drugs, incidence of anesthesia in connec-

Table 7. Symptoms (upper row) and signs (lower row) possibly indicative of polyneuropathy in the exposed and control groups, as recorded in the standardized questionnaire and standardized neurological examination.

Exposed group (n=30)	Final control group (n=30)	Statistical difference (p)	Original control group (n=30)	Statistical difference (p)	Original and final control groups (n=60)	Statistical difference (p)
12	5	0.086	7		12	0.077
18	15		12		27	

tion with surgery, meningo-encephalitis, concussion with or without unconsciousness, sciatica, anemia, hypertension, hyperlipemia and serious systemic diseases), the observed frequency was of the same order of magnitude in the exposed and nonexposed groups. For the use of alcohol, however, a quantitative estimation revealed a higher monthly average consumption in the nonexposed than in the exposed group: 188 and 120 gm, respectively (p = 0.12).

Psychiatric interviews and ratings

Prevalence of mental symptoms. The evaluation of the psychiatric interview yielded a score estimating the amount of symptoms in the individual. The individual scores and the difference in the scores of the matched subjects are given in table 8. The score for the exposed workers was larger than that of the controls in 20 of 30 pairs. In two pairs the scores were equal. In eight pairs the control's score was larger than that of the exposed worker. The average score of the exposed group ($\bar{S}_E = 8.6$) exceeded that of the control group ($\bar{S}_C = 5.6$). The average difference was $\bar{d} = 3.0$.

The results show that exposed individuals have more psychiatric symptoms than the controls (sign test, p < 0.01, Mann-Whitney p < 0.01, Student's *t* p < 0.001).

The presented scores show the sum of all rated items (1—37). The total score for each item in the exposed and control groups is illustrated in fig. 5. It can be seen that some of the items do not contribute to the difference mentioned. The relative importance of the rated items was tested after the items were grouped into

five categories (fig. 5). Neurasthenic symptoms (items 3—5, 13—16, 18, 19, 22 and 23) showed the largest differentiation between the groups, followed by neurotic disturbances (items 8, 10, 11 and 12). Dysmnesia,

Table 8. Individual scores of mental symptoms for exposed workers (S_E) and controls (S_C). Difference in score indicated by d.

Pair no.	S_E	S_C	d
1	4.00	3.00	1.00
2	15.25	7.25	8.00
3	6.25	2.50	3.75
4	5.25	5.00	0.25
5	3.25	5.75	—2.50
6	4.50	5.50	—1.00
7	3.75	4.00	—0.25
8	9.00	6.50	2.50
9	8.50	13.50	—5.00
10	9.25	10.50	—1.25
11	4.50	3.50	1.00
12	4.75	0	4.75
13	11.00	7.50	3.50
14	10.25	3.00	7.25
15	5.00	6.25	—1.25
16	15.50	12.50	3.00
17	15.50	11.50	4.00
18	4.00	3.50	0.50
19	8.25	4.00	4.25
20	15.00	3.75	11.25
21	3.25	3.25	0
22	12.50	5.00	7.50
23	3.75	1.75	2.00
24	21.25	1.00	20.25
25	1.25	4.50	—3.25
26	6.00	6.50	—0.50
27	6.75	3.75	3.00
28	9.75	9.75	0
29	18.75	8.25	10.50
30	10.50	6.00	4.50
$\bar{S}_E = 8.6$	$\bar{S}_C = 5.6$		$\bar{d} = 3.0$

211

somatic symptoms, and items depending on personality traits, did not differentiate between the examined groups.

Medical history. The evaluation of the medical history yielded an index of liability for the psychiatric symptoms of each subject. No difference in the liability index of the examinees was found in 17 pairs. The exposed workers showed a lower index than the controls in nine pairs, whereas the reverse result was obtained for four pairs. Thus, no preponderance of individuals with a high liability for psychiatric symptoms (i.e., indices) was found in the exposed group. No relation between the individual's score and index was found.

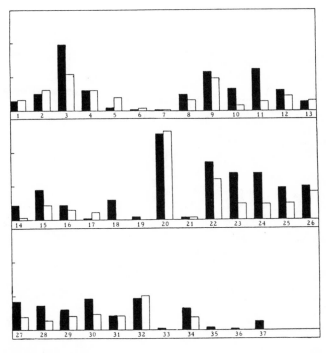

Category no. and items	P	NP
1. Temperament, 1, 2, 6, 7, 35, 36	—	—
2. Somatic, 26, 28—34, 37, 17, 20, 21	—	—
3. Dysmnesia, 24, 25, 27	—	*
4. Neurotic, 8, 10, 11, 12	*	—
5. Neurasthenic, 3—5, 13—16, 18, 19, 22, 23	*	*

* p < 0.05

Fig. 5. Histogram of relative prevalence of different mental symptoms in the exposed (dark) and control groups (white). The items are numbered according to the listing in the appendix. Below the histogram are the results of the test of difference between the examined groups after the items (1—37) were classified into five categories. Item numbers are listed for each category. Results of the nonparametric (NP) and parametric tests (P) are indicated.

Psychological tests

The statistical analysis of the results of the performance tests was carried out as follows.

The sampling distribution of each performance variable was inspected with regard to normality. The significance of the differences between the exposed and nonexposed groups was tested with regard to the homogeneity of variances and to the mean value of each performance variable. The statistical tests were performed both under the assumption that the groups were independent samples and under the assumption that the groups were correlated samples (matched pair). Furthermore, a nonparametric test (Mann-Whitney U test) was applied to the data as a complement to the parametric tests. Two of the performance tests, Simple RT and Bourdon—Wiersma, could be analyzed for performance changes during the test period. For these two tests a two-way analysis of variance with repeated measurements was employed in which the

Table 9. Mean values (M) and standard deviations (SD) of the performance on different tests for the exposed and nonexposed groups.

Test	Exposed group		Nonexposed group	
	M	SD	M	SD
Simple RT (ms)				
Speed (M)	275	44	270	31
Regularity (SD)	19	9	19	13
RT Addition (s)				
Speed (M)	3.0	1.0	2.8	1.0
Regularity (SD)	1.2	0.4	1.0	0.4
Memory test: recognition				
Number correct	29.5	2.9	28.7	4.0
Memory test: reproduction				
Criterion 1[a]	8.0	3.3	8.2	3.7
Criterion 2[b]	53.8	10.2	54.6	10.3
Manual dexterity: Santa Ana				
Right	49.1	7.9	48.0	6.8
Left	47.0	7.4	45.2	6.1
Coordination	31.6	8.4	31.3	7.5
Perceptual speed: Bourdon-Wiersma (s)				
Mean time per line	14.0	2.5	12.8	2.1

[a] Criterion 1: Number of correctly reproduced combinations.
[b] Criterion 2: Number of correctly reproduced elements.

Table 10. Analysis of variance of mean reaction time in the Simple RT test and mean time per line in the Bourdon-Wiersma test.

Source	Simple RT			Bourdon-Wiersma		
	df	Mean square	F	df	Mean square	F
G (groups)	1	3,256.07	0.26	1	312.53	4.16*
I (individuals)	58	12,544.56		58	75.22	
T (1-min time blocks)	8	5,179.46	13.10***	4	70.18	22.21***
G x T	8	461.86	1.17	4	3.94	1.25
Residual I	464	395.48		232	3.16	
T linear trend	1	39,441.96	55.79***	1	200.91	33.57***
G x T linear trend	1	2,704.00	3.82	1	7.22	1.21
Residual II	58	707.02		58	5.98	

* $p < 0.05$; ** $p < 0.01$; *** $p < 0.001$.

trend of the performance changes over time was introduced as a further source of variation.

Table 9 presents the mean values and standard deviations of the performance variables for the exposed and nonexposed groups. The analysis of variance for the two memory tests and for the manual dexterity test yielded F ratios that were very close to unity, and thus no differences in the performance of the groups were revealed on these tests. In the RT Addition test the exposed group had a longer reaction time and a greater intraindividual variation on the average than the nonexposed group. However, only the latter performance variable, i.e., the regularity of the individual performance on the test, resulted in a statistically significant difference in the analysis of variance [F $(1,59) = 4.12$; $p < 0.05$] and in the Mann—Whitney U test ($z = 2.22$; $p < 0.05$). The analysis of the results of the Simple RT test and the Bourdon—Wiersma test are given in table 10. There was no significant difference between the groups with regard to mean reaction time in the Simple RT test. As expected from the use of this test in previous research (17, 18), there was a significant linear trend over the time blocks [F $(1,58) = 55.79$; $p < 0.001$], i.e., the subjects' reaction time increased linearly during the test period. The analysis also showed that the linear change of the reaction time over the time blocks interacted with the groups [F $(1,58) = 3.82$; $0.05 < p < 0.10$]. In other words, time on task appears to affect the exposed subjects more than the nonexposed (see fig 6).

The Bourdon—Wiersma test also showed a performance decrement over time, but no differences in this respect were found between the groups. However, the analysis showed a significant difference in the mean performance between the groups [F $(1,58) = 4.16$; $p < 0.05$], the performance of the exposed subjects being poorer. The same result was also obtained with the Mann—Whitney U test ($z = 2.17$; $p < 0.05$).

To summarize, in the psychological investigation differences between the groups were found in three of the tests which made high demands on attention and sensorimotor speed. The investigation did

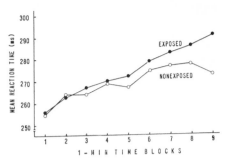

Fig. 6. Change in mean reaction time over time (1-min time blocks) for the exposed and nonexposed groups.

not reveal any difference between the groups in the tests concerning memory and manual dexterity function.

Neurophysiological examinations

(a) Electroencephalograms

The evaluation of the EEG recordings yielded the following results: At visual inspection all the EEGs were classified as normal except five in the exposed group and four in the control group. All abnormal findings in the exposed group were slight or moderate, episodic deviations within the frontotemporal areas with a left hemisphere dominance. The same type of abnormality was found also in the control group, except for one person who had a moderate abnormality within the right temporal region.

It is well-known, however, that the concept "normal EEG" of the clinical routine covers many types of EEG. It was the case also in this material. Despite the fact that the EEG recordings were judged as normal, sometimes their characters seemed to be widely different, as illustrated in fig. 2. Fig. 7 shows the visual ranking and the cumulative rank distribution. As can be seen, the control subjects tend to cluster to the left while the exposed ones gather at the right end. There is a significant difference between the two groups (Mann—Whitney, $z = 2.03$, $p < 0.05$).

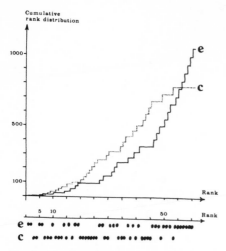

Cumulative
rank distribution

Rank

Rank

Fig. 7. Visual ranking and cumulative rank distribution of all EEGs. (e: exposed subjects; c: controls; N = 60 subjects)

Table 11 shows the means and standard deviations of the SPA parameters in the exposed and control groups for the leads P3—01 and P4—02. It can be seen that in comparison with the controls the exposed subjects showed a larger value for (a) the delta bandwidth (σ_δ) in P4—02, (b) the delta power (G_δ) bilaterally, (c) the alpha peak frequency (f_α) bilaterally, (d) the alpha bandwidth (σ_α) bilaterally, and (e)

the model-fit parameter in P4—02 and a smaller value of the alpha power (C_α) bilaterally. The differences were not significant. The exposed group, however, showed a larger interindividual variation (standard deviation) than the controls (F test, p < 0.05).

The ranking based on the different SPA parameters separately showed differences between the exposed and control subjects only for the alpha peak frequency (f_α), which was higher for the exposed subjects (Mann—Whitney, P3—01: p < 0.05 and P4—02: p < 0.08). The total ranking of all the subjects based on the sums of all parameter ranks of each subject showed no differences between the two groups. It is, however, interesting to note that out of the 16 subjects excluded from the SPA analysis, 10 were assigned visual ranks over 40 (maximum rank was 60). The result of the visual inspection was, however, not dependent on the 16 subjects excluded from the SPA processing. Both for 60 and 44 subjects differences were obtained between exposed and control subjects (Mann—Whitney, p < 0.05).

When the results of the SPA analysis were compared between the matched subjects and not between the examined groups as before, significant differences were obtained (analysis of variance, p < 0.05). Each parameter was tested within each matched couple. Fig. 8 shows the results of this study.

Table 11. Means and standard deviations of the SPA parameters of the exposed and control groups for the leads P3-01 and P4-02.

Lead	Subjects	Spectral parameters								Model fit
		σ_δ Hz	G_δ %	f_α Hz	σ_α Hz	G_α %	f_β Hz	σ_β Hz	G_β %	
P3-01	Controls									
	Mean	1.80	18	9.29	0.56	69	16.9	2.93	13	0.56
	SD	0.90	8	0.71	0.33	14	1.9	1.24	11	0.20
	Exposed									
	Mean	1.99	24	9.62	0.84	59	16.7	3.05	17	0.59
	SD	1.30	15	1.39	0.89	22	2.4	1.66	15	0.25
P4-02	Controls									
	Mean	1.74	21	9.39	0.55	65	16.3	3.54	14	0.51
	SD	1.33	11	0.73	0.29	15	2.3	1.66	10	0.20
	Exposed									
	Mean	2.05	28	9.72	0.88	56	16.3	3.84	16	0.61
	SD	1.40	18	1.40	0.91	22	2.5	1.98	15	0.22

(b) *Peripheral nervous system examinations*

Conduction velocities and compound nerve action potential amplitudes in peripheral nerves. Table 12 presents the ENeG findings for the exposed and nonexposed groups expressed as the mean differences between the matched subjects. The results showed that the NAP of the sural nerve was lower in the exposed than in the nonexposed group (p < 0.03). The SCV of the distal part (dse—S1 in fig. 4) of the ulnar nerve was slower in the nonexposed group when compared with that of the exposed group (p < 0.04) and, in addition, the MCV of the median nerve was slower in the nonexposed group (p < 0.06). The other parameters did not differentiate between the examined groups. (Only p values ≤ 0.10 in the table).

Sensation thresholds of vibration in the extremities. Table 13 presents the sensation thresholds of vibration for the exposed and nonexposed groups expressed as the mean differences between the matched subjects. The vibration thresholds were, on the average, constantly lower in the nonexposed group, although not statistically significantly lower.

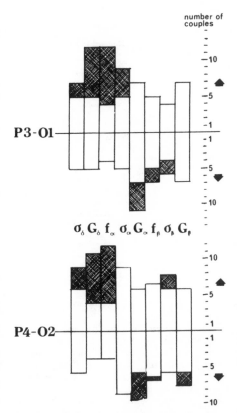

Fig. 8. Diagram showing the number of couples significantly differing in the SPA parameters. Upward column: Number of couples in which the exposed subject showed a significantly larger value than the control. Downward column: Number of couples in which the exposed subject showed a significantly smaller value than the control. For each parameter the difference in column sizes is indicated by a shadowed area.

DISCUSSION

Exposure

Exposure is a complex concept and includes parameters such as exposure intensity in terms of air concentrations of contaminant, duration of exposure, frequency of exposure at peak intensities, time lapse from first exposure to point of observation (latency period). Also, it may be subject to changes over long periods of time. Obviously, for an analysis of exposure in an epidemiologic context all such parameters would have to be considered. In view of the limited size of the studied groups, it has not been considered meaningful to divide the exposed group into subgroups according to these various parameters. The data at hand do not provide sufficient material for such an analysis.

It was, however, found that the exposure conditions in the work procedures studied were reasonably stable. It was observed that the variations of concentration around a mean value were not impressive and were slow and continuous. Therefore it was considered appropriate to describe the exposure to the factor under study as a calculated time-weighted average air concentration during work with relevant exposure. This entity is obviously an approximation. It is based on a limited number of measurements. The results are general-

Table 12. The ENeG findings for the exposed and nonexposed groups as expressed in the mean differences ($\overline{D}$) between the matched pairs. A positive mean difference indicates a smaller value for the exposed group.

ENeG parameters [a]	Number of matched pairs	$\overline{D}$	SD	Statistical differences (p)
N suralis				
NAP (S2—S1)	29	2.24	5.4	0.03
SCV (S2—S1)	26	− 1.23	6.0	
N peroneus				
MCV (S2—S1)	30	1.07	6.7	
N medianus				
NAP (dse—S1)	29	0.48	5.5	
SCV (dse—S1)	27	− 2.19	7.0	
NAP (S1—S2)	29	4.28	18.9	
SCV (S1—S2)	29	− 0.83	4.8	
MCV (S2—S1)	30	− 2.07	5.8	0.06
N ulnaris				
NAP (dse—S1)	30	0.50	5.1	
SCV (dse—S1)	29	− 3.45	8.5	0.04
CVSF (S2—S1)	30	1.40	5.9	
MCV (S2—S1)	30	− 0.43	6.3	

[a] See fig. 4 for an explanation of the points in parentheses.

Table 13. The sensation thresholds of vibration for the exposed and non-exposed groups expressed as the mean differences ($\overline{D}$) between the matched subjects. A positive mean difference indicates a higher vibration threshold value for the exposed group.

Vibration thresholds	Number of matched subjects	$\overline{D}$	SD	Statistical differences (p)
Carpal	30	0.19	0.74	
Tibial	30	0.22	4.91	
Tarsal	30	0.51	8.22	

ized to describe a typical exposure for categories of personnel. Thus, the results are not truly individually based in the sense that every member of the exposed group has not had his personal work situation subjected to measurements.

Medical history, standardized interviews and neurological examination

All but four subjects in the exposed group reported recurrent acute symptoms upon exposure to jet fuel vapors during work (dizziness, fatigue, headache, nausea, respiratory tract symptoms, palpitations and thoracic oppression). The incidence of neurasthenia, anxiety and/or mental depression during employment at the factory was significantly higher in the exposed than in the nonexposed groups as recorded in (a) the medical history and a standardized interview and (b) notations in the medical records of the factory health department when the employees had consulted, and been treated by, the plant physician for such symptoms. Among the neurasthenic symptoms fatigue, depressed mood, lack of initiative, and dizziness dominated, but significant differences were also found for palpitations and thoracic oppression, sleep disturbances, and headache. Also symptoms of gastritis and eye irritation were more frequent in the exposed than in the nonexposed groups (p < 0.1 and p < 0.01, respectively). As to the occurrence of symptoms and signs possibly indicative of polyneuropathy, trends towards higher prevalences were found in the exposed group.

In this part of the investigation it is worth emphasizing that the standardized interview and the information collected from the medical records of the factory health department yielded more or less identical results as to incidence of neuropsychiatric ill health among the examined. The fact that differences existed also when the medical records were scrutinized seems to rule out any bias of the observer as well as of the examined subjects. The increased incidence of symptoms in the exposed group is furthermore strengthened by comparisons with the original control group. It appears from the results that the original and final control group are very similar as regards incidence of symptoms.

Psychiatric interviews and ratings

After evaluation of the tapes all interviews were considered to be adequate for further analysis. No relation between reliability and amount of mental symptoms was noted. This finding suggests that overstatements by the interviewer or examinee did not significantly contribute to the results. In addition, there was good agreement between the first and second evaluation of the psychic symptoms by the interviewer, and also between independent raters with and without knowledge of the occupation of the examinee. Thus the evaluation does not seem to have been influenced by observer bias. No general inclination for complaining was found for the exposed group, since the difference between the examined groups was mainly due to neurasthenic symptoms.

The rating scale used to estimate the degree of the psychic symptoms was similar to that of another scale [CPRS (1)]. The CPRS was constructed to measure symptoms before and after pharmocological treatment of depressed and schizophrenic patients, and its reliability for this purpose was found to be high. In the present study, items of the CPRS concerning psychotic symptoms were replaced by other items concerning neurasthenia. The previously mentioned assessment is probably valid for the modification used since its structure was similar to that of the CPRS, which has also been used for volunteers without psychiatric disease (1).

It may be argued that the difference in psychic symptoms between the exposed and control groups is due to social or mental factors. However, no preponderance of large indices was found for the exposed group. Good agreement between independent index raters was observed. The validity of the index as a measure of probability for mental ill health has not been assessed. However, patients with a large amount of some previous somatic diseases and other items used are known to run a higher risk for the development of psychiatric symptoms. The result indicates that previous illness or constitutional factors did not contribute to the difference in amount of mental symptoms found between the groups.

The results thus show a higher prevalence of neurasthenic symptoms in workers exposed to jet fuel.

Psychological tests

The results of the examination using behavioral performance tests indicated a difference in performance capability between the groups. The exposed subjects had a greater irregularity of performance on a test of complex reaction time, a greater performance decrement over time in a simple reaction time task and poorer performance in a task of perceptual speed than the nonexposed subjects. No differences between the groups were found in two tests of memory functions and in a manual dexterity test.

When the psychological test results are interpreted, two circumstances should be noted which reduced the power of the statistical analysis of the results, i.e., the theoretical probability of finding statistically significant differences between the exposed and nonexposed groups of subjects. The first is that the analyses were based on relatively small samples. The second is that the examined groups were somewhat more heterogeneous with respect to education and, consequently, even with respect to mental performance capability than most of the groups examined in previous studies on the effects of long-term exposure to industrial solvents (17, 18, 22, 23, 36). This fact was confirmed by the comparison which could be made for two of the tests with the results of earlier studies. Thus, there was a greater interindividual variation of performance among the present workers than among previously studied groups. The fact that a statistically significant performance decrement among the exposed subjects only appears for some of the tests may to some extent be attributed to the circumstances and possibly to the differential sensitivity of different tests.

Neurophysiological examinations

Electroencephalograms. The appearance of abnormal EEGs in this material was 15 %, which is not remarkable in a clinically healthy population (11).

A ranking of the mainly normal EEGs of the control and exposed subjects yielded, however, a significant (p < 0.05) difference. The exposed group showed, on the average, a lower amplitude and a less observable rhythmic activity than the control group, which was more concentrated in the direction of the most "normal" end of the ranking scale.

Also the SPA yielded statistically significant differences between the two groups for some parameters. The results (significant or trends) were essentially in agreement with the visual ranking. It should be noted, however, that the SPA analysis concerned only the two single EEG leads which gave detailed information, while the visual inspection covered the whole EEG record.

The SPA changes essentially agree with what can be seen in an EEG of increasing degree of "abnormality" (27). It is, however, remarkable that the alpha peak frequency appears to be higher for the exposed subjects than for the controls. The physiological reason is unclear, but to some extent it is contradictory to an EEG change acutely induced. From studies of, for example, alcoholic intoxications, it is known that the frequency of the basic rhythm decreases due to the acute in-

fluence of alcohol (21). The chronic effect of organic solvents on the EEG is little known even if many authors report an increased amount of abnormal EEGs among subjects chronically exposed to organic solvents (42).

The SPA findings imply that the exposed group showed a lower degree of frequency stability of the EEG, as can be seen both from the model-fit parameter and the bandwidth (σ_α) of the alpha component, as well as from the comparison of the interindividual variability of the parameters. The alpha activity is regarded as a result of a thalamic control of the cortical activity (2). The harder the control, the larger the effect on the alpha activity, i.e., a higher value of the power G_α and a narrower banded (σ_α) alpha peak in the power spectrum. On the other hand, if the control vanishes completely, the whole signal acquires the character of low-pass filtered noise, i.e., only a delta component, G = 100 %. A small value of the bandwidth reflects a frequency-stable thalamic control. Therefore, the signal essentially has time-invariant spectral properties which yield a good model adaptation to the signal, i.e., a low value of the model-fit parameter. The SPA results might indicate that the effect of jet fuel is, among other things, an influence on the thalamic control of the cortical activity with an increased time variability, decreased frequency stability, and a less widespread control of the cortical neurons, i.e., an instability in the thalamocortical system.

Peripheral nervous system examinations. The results of the examination of the peripheral nerve functions with ENeG methods and vibration threshold determinations indicated differences between the groups in terms of (a) smaller NAPs of sensory nerves and an overrepresentation of higher vibration thresholds of the extremities in the exposed group and (b) slower SCVs and an overrepresentation of slower MCVs in the nonexposed group.

Clinical practice gives support to the view that measuring the NAPs of the peripheral part of sensory nerves is one of the most sensitive ENeG methods to detect early nerve function impairment of light degree. The method of measuring vibration thresholds is also considered sensitive for the diagnosis of such functional disturbance. With these facts in mind, one should recall the findings of smaller NAPs of sensory nerves in the exposed group and, on the average, a higher vibration threshold of the extremities. These findings agree with the results of the quantitive neurological examination. The slow conduction velocities in the ulnar (SCV) and median (MCV) nerves of the controls is an intriguing finding. The difference observed is probably not due to nerve damage however. The SCVs in these nerves are considered a late indicator of polyneuropathy, when compared to the NAP amplitudes, which were larger among the controls. The paradox may be explained by an occasional sampling of individuals with excellent peripheral nerve function in the exposed group. If so, the difference in NAP amplitude between the examined groups is an underestimate of the true impairment of nerve function in the exposed subjects.

GENERAL DISCUSSION

The present and the first study (32) on jet fuel exposed industrial workers show many similarities, e.g., more than two-thirds of the exposed subjects in both investigations stated acute symptoms repeatedly to occur in association with exposure to jet fuel vapors and about three-fourths had an incidence of neuropsychiatric ill health during the years of exposure. Also as to type of symptoms, there are similarities, dizziness being prominent among the acute symptoms and depressed mood, lack of initiative, dizziness, palpitations, chest oppression, sleep disturbances, and headache being frequent in a non-acute "syndrome" of neuropsychiatric health disturbance.

In the psychiatric interviews made independently of the aforementioned investigations, the exposed subjects showed an increased prevalence of neurasthenic symptoms when compared with the controls. The results of the psychiatric interviews thus are consistent with the evaluations on mental health made by non-psychiatrists and also with earlier descrip-

tions of chronic fuel intoxications, and the results of the first jet fuel study. The results also agree with those of a study on Swedish Air Force personnel exposed to jet fuel during aircraft servicing. The fuel exposure in this group was considerably less than that of the exposed group in the present study. Many subjects were exposed outdoors when refueling jet planes or at work with the planes in hangars. Only a slight fraction of the examined service men were exposed in the same order of magnitude as the exposed workers of the present study. All the service men were examined by psychiatrists, and the results showed that symptoms indicative of minor brain damage were more frequent among 142 highly exposed persons than among 105 persons with a low exposure to jet fuel (Jansson, Siwers and Antoniew, personal communication). Since the 247 examined were selected on the basis of exposure as well as subjective symptoms (Strandberg, personal communication), the result is probably an underestimate of the difference between the groups.

Petroleum-distilled fuels chemically consist of different organic hydrocarbons. Also the industrial solvents and some of the components in plastics and paints are hydrocarbons. In recent epidemiologic studies on house painters (4, 5, 6, 8) and car painters (25) occupationally exposed to such solvents, the results strongly indicated an increased risk for neuropsychiatric illness to result from long-term solvent exposure. In a case referent study using a regional register of disability pensions, a relative risk of 1.8 was found for the contraction of neuropsychiatric disease among workers exposed to solvents, e.g., painters, varnishers and carpetlayers, when compared to building workers without such exposure (6). In one of the previously mentioned studies (5), case reports were given on ten heavily exposed painters, who were found to have symptoms such as deterioration of memory, personality changes of asthenic or depressive-aggressive types and abnormal fatigue. The results led the authors to suggest a chronic psycho-organic syndrome in these patients.

The symptoms recorded among the jet fuel exposed workers fit well into such a syndrome. This is not surprising since the solvents and the fuels are both composed of organic hydrocarbons. It is sobering to realize, however, that neuropsychiatric ill health may develop at concentration levels well below that which, until recently, has been considered safe for exposed workers.

The reported effects on mental health cannot, furthermore, be disregarded as marginal or unimportant. In view of both risk source prevalence and the impact of disease in terms of personal suffering and social burden, exposure to organic hydrocarbons should be recognized as a cause of serious concern. We feel that it is advisable to reevaluate the basis of threshold limit values set for organic solvent products and — in so doing — give greater consideration than hitherto to the risk for long-term effects on the nervous system.

SUMMARY

Thirty jet fuel exposed workers selected according to exposure criteria and thirty nonexposed controls from a jet motor factory were examined with special reference to the nervous system by occupational hygiene physicians, psychiatrists, psychologists, and neurophysiologists. The controls were matched with respect to age, employment duration, and education. Among the exposed subjects the mean exposure duration was 17 years, and 300 mg/m^3 was calculated as a rough time-weighted average exposure level.

All but four subjects in the exposed group reported recurrent acute symptoms upon exposure to jet fuel vapors during work (dizziness, fatigue, headache, nausea, respiratory tract symptoms, palpitations, and thoracic oppression). The incidence of neurasthenia, anxiety and/or mental depression during employment at the factory was significantly higher in the exposed than in the nonexposed groups, as recorded in (a) the medical history and (b) a standardized interview and (b) notations in the medical records of the factory health department. Among the neurasthenic symptoms fatigue, depressed mood, lack of initiative, and dizziness dominated, but significant differences were also found for palpitations and thoracic oppression, sleep disturbances, and headache.

The prevalence of mental symptoms in the exposed and nonexposed groups was assessed from psychiatric interviews and ratings. The interviews were taped and later evaluated, with the use of rating scales, independently by the interviewing psychiatrist and another psychiatrist, who had no prior knowledge of the subject and did not know whether the subject was exposed or not. The correlation was found to be good between the ratings made by the interviewer and by the second psychiatrist. The evaluation of the interviews resulted in a score estimating the amount of symptoms in the individuals. Comparisons showed the score of the exposed workers to be larger than that of the controls in 20 of 30 pairs. In two pairs the scores were equal, and in eight pairs the control score was larger than that of the exposed worker. The average score of the exposed group exceeded that of the control group. The results show that exposed individuals have more psychiatric symptoms than the controls. Neurasthenic symptoms showed the largest differentiation between the groups, followed by neurotic disturbances.

As to occurrence of symptoms and signs possibly indicative of polyneuropathy, trends towards higher prevalences were found in the exposed group. The results of the examination of the peripheral nerve functions with electroneurographic methods and vibration threshold determinations indicated differences between the groups in terms of (a) smaller nerve action potentials of sensory nerves and a tendency toward higher vibration thresholds of the extremities in the exposed group and (b) slower sensory nerve conduction velocities and a tendency toward slower motor nerve conduction velocities in the nonexposed group.

In the psychological examination behavioral performance tests indicated a difference in performance capability between the groups. The exposed subjects had a greater irregularity of performance on a test of complex reaction time, a greater performance decrement over time in a simple reaction time task, and a poorer performance in a task of perceptual speed than the nonexposed subjects.

The EEGs recorded were considered clinically normal in almost all of the cases.

However, this normality concept includes many EEG varieties. Thus, when the EEGs were ranked as to configuration of alpha activity, significant differences were obtained between the two groups. The exposed group showed, on the average, a lower amplitude and a less observable rhythmic activity than the control group, the EEGs of which were clustered in the most "normal" end of the ranking scale. Also a computer assisted spectral parameter analysis of the EEGs resulted in significant differences between the groups.

To summarize, the present investigation revealed significant differences between the exposed and nonexposed groups for (a) incidence and prevalence of psychiatric symptoms, (b) psychological tests with load on attention and sensorimotor speed and (c) EEGs. In the selection of the control group it was ensured that the two groups were essentially equivalent except for exposure to jet fuel. It is concluded, therefore, that the differences found between the groups are probably related to exposure to jet fuel.

REFERENCES

1. ÅSBERG, M., MONTGOMERY, L., PERRIS, C., SCHALLING, D. and SEDVALL, G. The comprehensive psychopathological rating scale. *Acta psychiatr. scand. suppl.* (In press)
2. ANDERSEN, P. and ANDERSSON, S. A. *Physiological basis of the alpha rhythm.* Appleton-Century-Crofts, New York, N.Y. 1968. 235 p.
3. AXELSON, O. *Scandinavian meeting on health hazards in the use of solvents, April 15—16, 1975.* Department of Occupational Medicine, Regionsjukhuset, Örebro.
4. AXELSON, O., HANE, M. and HOGSTEDT, C. Case reports on chronic psycho-organic syndrome in house painters (In Swedish). *Läkartidningen* 73 (1976) 317—318.
5. AXELSON, O., HANE, M. and HOGSTEDT, C. Psychological function changes among house painters (II) (In Swedish). *Läkartidningen* 73 (1976) 319—321. [Summarized in: HANE, M., AXELSON, O., BLUME, J., HOGSTEDT, C., SUNDELL, L. and YDREBORG, B. Psychological function changes among house painters. *Scand. j. work environ. & health* 3 (1977) 91—99.]
6. AXELSON, O., HANE, M. and HOGSTEDT, C. Neuropsychiatric ill-health in workers exposed to solvents: A case-control study (In Swedish). *Läkartidningen* 73 (1976) 322—325.
7. BERGSTRÖM, J., LINDBLOM, U. and NORÉE, L.-O. Preservation of peripheral nerve function in severe uremia during treatment with low protein high caloric diet and surplus of essential amino acids. *Acta neurol. scand.* 51 (1975) 99—109.
8. BLUME, J., HANE, M., SUNDELL, L. and YDREBORG, B. Mental function changes among house painters (In Swedish). *Läkartidningen* 72 (1975) 702—706. [Summarized in: HANE, M., AXELSON, O., BLUME, J., HOGSTEDT, C., SUNDELL, L. and YDREBORG, B. Psychological function changes among house painters. *Scand. j. work environ. & health* 3 (1977) 91—99.]
9. BOURRET, J., VIALLIER, J., TOLOT, F. and ROBILLARD, J. Polynévrites par exposition simultanée au trichloroéthylène et à l'essence. *Rev. méd. suisse romande* 88 (1968) 173—181.
10. BUCHTHAL, F. and ROSENFALCK, A. Evoked potentials and conduction velocity in human sensory nerves. *Brain res.* 3 (1966) 1—122.
11. COBB, W. A. The normal adult EEG. In: D. HILL and G. PARR (eds.), *Electroencephalography.* Macdonald, London 1963, pp. 221—249.
12. CONTAMIN, F., GOULON, M. and MARGAIRAZ, A. Polynévrites observées chez des sujets utilisant comme moyen de chauffage des appareils a combustion catalytique de l'essence. *Rev. neurol.* 103 (1960) 341—354.
13. DAVIES, N. E. Jet fuel intoxication. *Aerosp. med.* 35 (1964) 481—482.
14. DRINKER, P. YAGLOW, C. P. and WARREN, M. F. The threshold toxicity of gasoline vapor. *J. ind. hyg. toxicol.* 25 (1943) 225—232.
15. DUVOIR, M., POLLET, L. and ARNOLDSON, M. La polynévrite benzinique exist-t-elle? *Soc. méd hôp.* (Paris) 54 (1938) 359—369.
16. FELIX, D. *Vierteljahrschr. f. Öffentl. Ges. Pfl.* (1872)
17. GAMBERALE, F., ANNWALL, G. and ANSHELM OLSON, B. Exposure to trichloroethylene: III. Psychological functions. *Scand. j. work environ. & health* 2 (1976) 220—224.
18. GAMBERALE, F., LISPER, H. O. and ANSHELM OLSON, B. The effects of styrene vapour on the reaction time of workers in the plastic boat industry. In: M. HORVÁTH (ed.), *Adverse effects of environmental chemicals and psychotropic drugs* (vol. 2). Elsevier Scientific Publishing Co., Amsterdam and New York, N. Y. 1976, pp. 135—148.
19. GERARDE, H. W. Aliphatic hydrocarbons. In: F. A. PATTY (ed.), *Industrial hygiene and toxicology* (2nd ed., vol II). Interscience, New York, N. Y. 1962, pp. 1195—1205.
20. GOODGOLD, J. and EBERSTEIN, A. *Electrodiagnosis of neuromuscular diseases.* The Williams and Wilkins Company, Baltimore, Md. 1972.
21. GREEN, R. L. The electroencephalogram in alcoholism, toxic psychoses and infection.

In: W. P. WILSON (ed.), *Applications of electroencephalography in psychiatry*, Duke University Press, Durham, N. C. 1965, pp. 123—139.

22. HÄNNINEN, H. Psychological picture of manifest and latent carbon disulfide poisoning. *Br. j. ind. med.* 28 (1971) 374.

23. HÄNNINEN, H., ESKELINEN, M. A., HUSMAN, K. and NURMINEN, M. Behavioral effects of long-term exposure to a mixture of organic solvents. *Scand. j. work environ, & health* 2 (1976) 240—255.

24. HAYHURST, E. R. Poisoning by petroleum distillates. *Ind. med.* 5 (1936) 53—63.

25. HUSMAN, K. R., SEPPÄLÄINEN, A. M., KARLI, P., HERNBERG, S. and ESKELINEN, L. The neurological and psychic effects of solvents on car painters. *18th International Congress on Occupational Health, Brighton, England, 14—20 September,* 1975

26. ISAKSSON, A. *The operating system for computer analysis of EEG at the Karolinska Hospital, Stockholm* (Technical report no. 85). Department of Telecommunication Theory, Royal Institute of Technology, Stockholm 1974.

27. ISAKSSON, A. and WENNBERG, A. Visual evaluation and computer analysis of the EEG — A comparison. *Electroenceph. clin. neurophysiol.* 38 (1975) 79—86.

28. JACOBZINER, H. and RAYBIN, H. W. Kerosene and other petroleum distillate poisonings. *N. y. state j. med.* 63 (1963) 3428—3430.

29. KNAVE, B., GOLDBERG, J. M., PERSSON, H. E. and WILDT, K. *Chronic exposure to lead: II. A. neurological and neurophysiological health investigation in a storage battery factory.* (Arbete och hälsa no. 4). Arbetarskyddsverket, Stockholm 1975. 14 p.

30. KNAVE, B., KOLMODIN-HEDMAN, B., PERSSON, H. E. and GOLDBERG, J. M. Chronic exposure to carbon disulfide: Effects on occupationally exposed workers with special reference to the nervous system. *Work-environ.-health* 11 (1974) 49—58.

31. KNAVE, B., KOLMODIN-HEDMAN, B., PERSSON, H. E. and GOLDBERG, J. M. *Chronic CS₂-poisoning: Effects on the nervous system of occupationally exposed workers.* (Arbete och hälsa no. 2). Arbetarskyddsverket, Stockholm 1973. 20 p.

32. KNAVE, B., PERSSON, H. E., GOLDBERG, J. M. and WESTERHOLM, P. Long-term exposure to jet fuel: An investigation on occupationally exposed workers with special reference to the nervous system. *Scand. j. work environ. & health* 2 (1976) 152—164.

33. KNAVE, B., PERSSON, H. E., GOLDBERG, J. M. and WESTERHOLM, P. Long-term exposure to jet fuel: An investigation on occupationally exposed workers with special reference to the nervous system. In:

M. HORVÁTH (ed.), *Adverse effects of environmental chemicals and psychotropic drugs: Neurophysiological and behavioural tests* (vol. 2). Elsevier Scientific Publishing Company, Amsterdam 1976, pp. 149—155.

34. LIÈVRE, J. A., BÉNICHOU, C. and DESROY, M. Polynévrites provoquées par le chauffage à catalyse. *Soc. méd. hôp.* (Paris) 118 (1967) 91—99.

35. LINDGREN, B. W. *Statistical theory* (2nd ed.). Macmillan Company, New York, N. Y. 1968.

36. LINDSTRÖM, K. Psychological performances of workers exposed to various solvents. *Work-environ.-health* 10 (1973) 151—155.

37. LUDIN, H. P. *Praktische elektromyographie.* Ferdinand Enke Verlag, Stuttgart 1976.

38. MACHLE, W. Gasoline intoxication. *J. am. med. assoc.* 117 (1941) 1965—1971.

39. OETTINGEN, W. F. VON. Toxicity and potential dangers of aliphatic and aromatic hydrocarbons. *U. s. public health bull.* 225 (1940) 43—65.

40. PERSSON, H. E., KNAVE, B., GOLDBERG, J. M. and JOHANSSON, B. *Chronic exposure to lead: IV. A neurological and neurophysiological health investigation in a heavy metals industry.* (Arbete och hälsa). Arbetarskyddsverket, Stockholm. (In press)

41. POTTS, C. S. A case of probable encephalitis due to the inhalation of the fumes of gasoline. *J. nerv. ment. dis.* 42 (1915) 24—27.

42. SEPPÄLÄINEN, A. M. Applications of neurophysiological methods in occupational medicine: A review. *Scand. j. work environ. health* 1 (1975) 1—14.

43. SEPPÄLÄINEN, A.-M. and HERNBERG, S. Sensitive technique for detecting subclinical lead neuropathy. *Br. j. ind. med.* 29 (1972) 443—449.

44. SPENCER, O. M. The effect of gasoline fumes on dispensary attendance and output in a group of workers. *Public health rep.* 37 (1922) 2291—2307.

45. STERNER, J. H. Study of hazards in spray painting with gasoline as a diluent. *J. ind. hyg. toxicol.* 23 (1941) 437—447.

46. TAKEUCHI, Y., MABUCHI, C. and TAGAKI, S. Polyneuropathy caused by petroleum benzine. *Int. Arch. Arbeitsmed.* 34 (1975) 185—197.

47. VIGDORTSCHIK, N. A. The problems of chronic action of benzine on the organism. *Zentralbl. Gewerbehyg. Unfallverhüt.* 20 (1933) 219.

48. WENNBERG, A. and ZETTERBERG, L. H. Application of a computer-based model for EEG analysis. *Electroenceph. clin. neurophysiol.* 31 (1971) 457—468.

49. ZETTERBERG, L. H. Estimation of parameters for a linear difference equation with application to EEG-analysis. *Math. biosci.* 5 (1969) 227—275.

Received for publication: 28 June 1977

APPENDIX

Contents of the interview and the items used for rating the liability for mental symptoms (A), the items used for rating different mental symptoms (B), and the principal steps of the scale used for rating the two sets of items (C)

A.
1. Hereditary factors
2. Factors in previous mental health
3. Factors in previous physical health
4. Way of life
5. Miscellaneous

B.
1. Sadness
2. Elation
3. Inner tension
4. Hostile feelings
5. Inability to feel
6. Pessimistic thoughts
7. Suicidal thoughts
8. Hypochondriasis
9. Worrying over trifles
10. Compulsive thoughts
11. Phobias
12. Rituals
13. Indecision
14. Inertia
15. Fatigability
16. Concentration
17. Reduced appetite
18. Reduced sleep
19. Increased sleep
20. Reduced sexual interest
21. Increased sexual interest
22. Aches and pains
23. Muscular tension
24. Recent memory
25. Long-term memory
26. Headache
27. Learning
28. Vertigo and dizziness
29. Chest oppression
30. Heart symptoms
31. Nausea
32. Gastrointestinal disturbances
33. Tics
34. Stimulus sensitivity
35. Mood lability
36. Impulsivity
37. Precision movements

C.
0.   None: Absence of symptoms (load)
0.5
1.   Minimal symptoms (load)
1.5
2.   Moderate and definite symptoms (load)
2.5
3.   Severe or incapacitating symptoms (load)

225

Scand j work environ health 6 (1980) 239—273

Exposure to organic solvents

A cross-sectional epidemiologic investigation on occupationally exposed car and industrial spray painters with special reference to the nervous system

by Stig-Arne Elofsson, PhD,[1] Francesco Gamberale PhD,[2] Tomas Hindmarsh, MD,[3] Anders Iregren, MSc,[2] Anders Isaksson, DSc,[4] Inger Johnsson, MSc,[5] Bengt Knave, MD,[2] Eva Lydahl, MD,[5] Per Mindus, MD,[5] Hans E Persson, MD,[5] Bo Philipson, MD,[5] Maria Steby, MSc,[2] Göran Struwe, MD,[5] Erik Söderman, MSc,[2] Arne Wennberg, MD,[2] Lennart Widén, MD [5]

ELOFSSON S-A, GAMBERALE F, HINDMARSH T, IREGREN A, ISAKSSON A, JOHNSSON I, KNAVE B, LYDAHL E, MINDUS P, PERSSON HE, PHILIPSON B, STEBY M, STRUWE G, SÖDERMAN E, WENNBERG A, WIDÉN L. Exposure to organic solvents: A cross-sectional epidemiologic investigation on occupationally exposed car and industrial spray painters with special reference to the nervous system. *Scand j work environ health* 6 (1980) 239—273. In the present epidemiologic study 80 car or industrial spray painters with long-term low level exposure to organic solvents were examined and compared with two matched reference groups of nonexposed industrial workers (80 persons in each group). The aim of the study was to investigate the possible effects of the solvent exposure on health. The investigation included psychiatric interviews, psychometric tests, neurological, neurophysiological and ophthalmologic examinations, and computed tomography of the brain. The painters' previous and present exposure was carefully assessed by interviews and on-the-job measurements both at modern places of work and in a reconstructed model of a workshop from 1955. On the basis of the psychiatric interviews the psychiatric symptoms were rated according to a specially designed scale of 46 different items, graded in seven steps of increasing severity. The psychological performance was assessed by a battery of 18 tests. The neurological and neurophysiological examinations comprised visual evoked responses (VER), electroencephalography (EEG), and computerized EEG analysis (SPA) for the central nervous system and eléctroneurography (ENeG), the estimation of vibration sense thresholds, and a quantified neurological examination for the peripheral nervous system. The ophthamologic examination concentrated on the condition of the lens. Statistically significant differences between the exposed individuals and referents were found for psychiatric items indicative of a slight cerebral lesion (ie, a neurasthenic syndrome). The psychometric tests revealed statistically significant differences between the groups with respect to reaction time, manual dexterity, perceptual speed, and short-term memory. No differences were found with respect to performance on verbal, spatial, and reasoning tests. Significant differences between the groups were also found for the majority of the neurophysiological parameters measuring peripheral nerve functions, the most pronounced occurring in the long, sensory fibers. Moreover EEG and VER showed some differences between the groups, as did the results of the ophthalmologic examination and the computed tomography. Finally, it should be emphasized that the exposure levels, as measured at modern places of work and in the reconstructed workshop from 1955, were found to be considerably lower than the valid threshold limit values in Sweden.

Key terms: computed brain tomography, epidemiology, lens opacities, neurasthenia, neurotoxicology, polyneuropathy.

[1] Department of Statistics, University of Stockholm, Sweden.
[2] National Board of Occupational Safety and Health, Solna, Sweden.
[3] Danderyd Hospital, Danderyd, Sweden.
[4] The Swedish County Council, Stockholm, Sweden.
[5] Karolinska Hospital, Stockholm, Sweden.

Reprint requests to: Dr Bengt Knave, National Board of Occupational Safety and Health, S-171 84 Solna, Sweden.

0355-3140/80/040239-35

The acute, narcotic effect of occupational exposure to organic solvents has long been known (14); it has also recently been the subject of systematic and experimental studies (2, 26). In the early literature, there are many case reports of exposed workers with chronic symptoms that indicate a more or less persistent injurious effect on the nervous system (59). A neurasthenic syndrome including abnormal fatigue, concentration difficulties, memory impairment, general irritability, and alcohol intolerance was described already at the end of the 19th century for workers exposed to carbon disulfide and during the first decades of the 20th century for workers exposed to other solvents, ie, trichloroethylene, xylene, terpentine. During 1930—1960 several investigations were published on groups of exposed workers.

The proportion of persons with neurasthenic symptoms was often about 50 %. The values of these results are however limited by the lack of reference material.

It is a fair assumption that the level of exposure was considerably higher during the 19th and the beginning of the 20th century than it is today, when the awareness of the work environment and its health hazards has gradually increased. Too, immediately after the Second World War, the levels of exposure were probably higher than they are today. This assumption is strengthened by the exposure determinations made in connection with a health examination of 50 trichloroethylene-exposed industrial workers (34). In this study in which 34 % of those examined showed signs of a chronic psychoorganic syndrome, the average exposure level was

Table 1. Levels of solvent exposure during different work phases of car and industrial painting. (Number of samples in parenthesis)

Solvents	Exposure, expressed as the hygienic effect[a] of the workplace						Threshold limit value (mg/m³)
	Spray painting		Grinding, filling, masking, assembling		Color mixing, degreasing, cleaning		
	Car painting	Industrial painting	Car painting	Industrial painting	Car painting	Industrial painting	
Total	0.30 (106)	0.24 (243)	0.15 (218)	0.12 (61)	0.28 (137)	0.24 (136)	
Aromates							
Toluene	0.13 (106)	0.05 (243)	0.05 (218)	0.04 (61)	0.13 (137)	0.04 (136)	300
Xylene	0.04 (106)	0.03 (243)	0.01 (218)	0.01 (61)	0.02 (127)	0.02 (136)	350
Styrene	0.04 (28)	0.01 (45)	0.02 (97)	0.08 (18)	0.08 (28)	0.01 (22)	170
Alcohols							
Ethanol	0.01 (51)	0.01 (136)	0.01 (96)	<0.01 (35)	0.01 (60)	<0.01 (51)	1,900
Butanol	0.04 (47)	0.01 (218)	0.01 (103)	— (0)	0.03 (59)	0.02 (22)	150
Propanol	— (0)	<0.01 (58)	— (0)	<0.01 (21)	— (0)	— (0)	500
Ketones							
Acetone	— (0)	<0.01 (45)	— (0)	<0.01 (18)	— (0)	<0.01 (58)	1,200
Methyl ethyl ketone	0.06 (15)	<0.01 (150)	<0.01 (22)	<0.01 (32)	<0.05 (18)	0.01 (58)	440
Methyl-n-butyl ketone	0.25 (6)	0.02 (92)	— (0)	0.05 (18)	— (0)	0.02 (58)	20
Methyl-iso-butyl ketone	0.01 (34)	0.03 (200)	<0.04 (105)	0.01 (53)	<0.01 (52)	0.01 (114)	210
Ethyl amyl ketone	— (0)	0.03 (43)	0.02 (32)	— (0)	— (0)	0.02 (11)	130[ᵇ]
Acetates							
Methyl acetate	— (0)	0.02 (82)	<0.01 (32)	— (0)	— (0)	0.03 (22)	610[b]
Butyl acetate	0.01 (49)	0.01 (59)	<0.01 (163)	<0.01 (26)	0.01 (82)	0.01 (5)	710
Ethyl acetate	0.01 (106)	<0.01 (198)	0.01 (218)	<0.01 (43)	0.01 (137)	<0.01 (78)	1,100
Iso-amyl acetate	<0.01 (27)	0.01 (54)	<0.01 (32)	— (0)	0.01 (11)	<0.01 (22)	525
Chlorinated hydrocarbons							
Methylene chloride	0.11 (8)	0.03 (45)	0.09 (22)	0.02 (18)	0.07 (30)	0.04 (58)	250
Trichloroethane	— (0)	— (0)	— (0)	— (0)	— (0)	0.57 (11)	380
Trichloroethylene	0.05 (15)	0.13 (161)	— (0)	0.06 (18)	0.04 (11)	0.16 (69)	105
Others							
White spirit[c]	0.17 (28)	— (0)	0.22 (42)	— (0)	0.16 (51)	— (0)	500
Percentage of the total work time	15	60	60	20	25	20	

[a] The hygienic effect of a compound is defined as the ratio between the actual amount of the compound and its threshold limit value. For an estimation of the total or additive hygienic effect caused by several compounds with similar effects, when specific information about synergism does not exist, the calculation $C_1/G_1 + C_2/G_2 + C_3/G_3 + \ldots + C_n/G_n$ is carried out, where C = the concentration of the compound and G = the threshold limit value of the compound. To avoid the exceeding of the total hygienic threshold limit value, the sum of the ratios are not allowed to exceed the value 1 (61).

[b] Swedish threshold limit value does not exist. The value used here is taken from the American Conference of Governmental Industrial Hygienists (5).

[c] 17—22 % by weight.

40 ppm, ie, twice as high as the present Swedish threshold limit value (TLV).

Around 1975 the first results from epidemiologic studies on solvent-exposed workers were published in Finland and Sweden. In investigations on industrial workers (52), house painters (9, 39), car painters (38, 75), and jet fuel-exposed workers (44, 45), the results indicated effects on the central and peripheral nervous systems as a result of long-term but relatively low exposure to different organic solvents. When the present study was initiated, the preliminary results of the epidemiologic studies had attracted great attention from both the general public and scientists, and uncertainty reigned as to whether or not the permitted limits could be regarded as adequate.

Against this background, it was considered pertinent to carry out an extensive epidemiologic investigation of a homogeneous group of workers with long-term occupational exposure to organic solvents. In consensus with representatives of employers and trade unions, we decided that car and industrial painters were an appropriate group. After a thorough occupational hygienic exposure determination, 80 spray painters, evenly distributed as to age, were selected as the exposed group and two groups of 80 age- and education-matched workers from the electronics industry were selected as reference groups. A total of 240 workers were then examined with psychiatric, psychological, neurological, neurophysiological, neuroradiological and ophthalmological methods.

EXPOSURE

At the workshops covered in this investigation the painting of repaired cars and new industrial products took place. The surface treatment also included degreasing, grinding, filling, assembling/reassembling of the treated objects, masking of untreated surfaces, and color mixing; the workers also had to clean tools and soiled skin. Analyses of 20 workshops showed that the relative time distribution of different work phases was somewhat different for the two groups of painters (table 1). The industrial painters primarily spray painted, while grinding, filling and masking occupied most of the work time of the car painters.

The chemical hazards of painters consist of solvents, dust, and metals. Exposure to solvents is the most frequent during color mixing, degreasing, cleaning, and the evaporation from painted surfaces and open cans containing solvents. There is less exposure during the spray painting itself because this work phase is mostly carried out in ventilated spraying boxes and with the use of protective respiratory equipment. However, as shown in table 1, a much larger proportion of the work time of industrial painters is spent on operations involving direct contact with solvents (spray painting, color mixing, degreasing, cleaning) than that of car painters. The workshop studies have shown that industrial painters are exposed to a higher degree than car painters during evaporation from painted objects which are air-dried on the work premises.

Painters are always exposed to several solvents. A total of about 20 different solvents has been found during personal sampling measurements (fig 1), and often the painters are exposed simultaneously to eight to ten substances (49, 81). Table 1 gives a quantitative description of this mixture of solvents in relation to the Swedish TLV. (The description is given as the so-called hygienic effect, see note a of table 1). Toluene is the solvent most frequently occurring in the highest concentration, but high concentrations of xylene, trichloroethylene, and white spirit are also common. The summarized (additive) hygienic effect varies for the different work phases between 0.12 and 0.30. If the time used for each work phase is taken into consideration, a time-weighted average of 0.21 is obtained for the hygienic effect of both car and industrial painting.

The exposure to dust, lead, and chromium is shown in fig 1. Lead is of special interest because of its well-documented neurotoxic effects. The table gives the range of hygienic effects, as well as the mean and the median exposure values, for spray painting and grinding. As a negligible exposure to dust occurs during color mixing, degreasing, and cleaning, no measurements of dust exposure were carried out during these work phases. The range of the exposure values show that the hygien-

Compound and work phase	Threshold limit value (mg/m³)	Exposure expressed as hygienic effect					
		Mean value	Median value	Measured value			
				0.01	0.1	1	10
Dust	10						
Spray painting							
Car painting		0.7	0.7				
Industrial painting		0.2	0.1				
Grinding							
Car painting		0.3	0.2				
Industrial painting		0.3	0.1				
Lead	0.1						
Spray painting							
Car painting		1.1	0.4				
Industrial painting		0.4	0.1				
Grinding							
Car painting		0.2	0.1				
Industrial painting		0.1	0.1				
Chromium	0.02						
Spray painting							
Car painting		1.3	0.7				
Industrial painting		0.5	0.3				
Grinding							
Car painting		0.3	0.2				
Industrial painting		0.3	0.2				

Fig 1. Exposure to dust, lead and chromium during spray painting and grinding. For each occasion of sampling the limits of detection were determined by the amount of sampled dust. For cases in which an exact measurement could not be obtained, a limit was determined which was not exceeded by the real value. In the figure this limit is represented by an unfilled circle. Exact measurements are represented by filled circles.

ic effects are sometimes exceeded during the spray painting of cars and, sporadically, also during the spray painting of industrial components. Generally, during grinding, the exposure level is lower. Calculations of the mean values give higher exposure values than the corresponding median value for all work phases.

The work conditions described are currently valid, ie, the exposure measurements performed in 1975—1977 can be considered to represent the work conditions of the 1960s and 1970s. As many of the painters included in this investigation have been occupationally exposed prior to this period and as the development of chronic effects in the nervous system due to long-term exposure can be suspected, a car-painting workshop was constructed to simulate the work conditions of 1955, and exposure measurements of solvents and lead were carried out. The results from this study have been presented separately (82). The main results are presented in table 2. It appears from the table that the exposure levels of solvents were higher for the simulated model than for the occasions of measurement in 1975—1977, especially with respect to the presence of benzene (additive effect 0.80). It should be observed that the TLV of benzene is low (15 mg/m³) in comparison to that of the other solvents referred to in this report, primarily because of its known effects on bone marrow and the blood and because of the fact that the absolute concentration in the air is therefore relatively low. However, the exposure levels in 1955 fell below the TLV for the solvents used today.

The values measured for lead showed a large variation depending upon the color used during the spray painting. Low concentrations of lead were measured during the use of all the colors except red. The exposure level with red exceeded the lead TLV 70-fold. It should be mentioned that the actual exposure to lead during the spray painting was reduced by the use of individual protective equipment.

EXAMINED GROUPS

Exposed group

The selection of the exposed group (N = 80) was based on information collected during semistructured individual interviews with 156 male Swedish spray painters (95 car painters and 61 industrial painters). During the interview they were questioned about occupational history, special attention being given to the intensity and duration of the exposure, the use of personal protective equipment, and the work conditions in each workshop. Furthermore, information was collected concerning environmental background and education. So that age would not become a confounding factor, the exposed group was stratified into five age groups, $\geq$ 25, 26—35, 36—45, 46—55 and 56—65 a. Each group consisted of eight car painters and eight industrial painters.

During the selection of the examined group, the exposed workers in each age group were ranked according to two indices of exposure. The indices were based on the period of employment during which exposure to solvents occurred (t), estimation of the intensity of the exposure on a three-grade scale (i), the use of personal protective equipment (p), and the work conditions in each workshop (c). The two exposure indices were calculated as t (i $\times$ p $\times$ c) and t (i + p + c).

Individuals ranking highly on both exposure indices were selected to be included in the exposed group. When the exposure indices did not give the same ranking or when the indices did not sufficiently differentiate the examined persons, the selection was made after careful comparisons between the data on occupational history. Thirteen persons, four belonging to the age group of car painters < 25 a and nine being randomly distributed among the other groups, refused to participate in the medical examinations and were replaced by the next person on the ranking list.

Reference groups

This investigation was planned as a cross-sectional epidemiologic study. The aim was to compare the spray painters to biologically and socially similar persons without exposure to organic solvents. The intention was therefore to select a group of nonexposed male Swedish industrial workers of the same age, type of work, and level of education as the spray painters. The selections were made in cooperation with representatives of the unions, managements, and health departments of the plants.

Reference group 1

Methods of selection. The first reference group was taken from an electronics plant in Stockholm. The selections were made from departments with (i) high- and (ii) low-qualified work. There was no individual classification of the education and description of occupation of the employees. Because the enterprise wished to avoid possible production disturbances, the reference group could not be chosen freely from all the employees, as some departments essential from the production point of view were not accessible to the investigators.

Table 2. Exposure to solvents and dust during car painting in 1975 and during simulated work conditions of car painting in 1955. — Distribution of time between different work phases.

Year	Exposure (hygienic effect)/percentage of the total time spent in different work phases		
	Spray painting	Grinding, filling, masking, assembling	Color mixing, degreasing, cleaning
1975			
Solvents	0.30/15	0.15/60	0.
Lead	0.40/15	0.10/60	
1955			
Solvents	0.40 (0.80 a) /20	—/65	0.25/15
Lead	0.0/20 (black)	—/65	—/15
	0.2/20 (green)		
	0.2/20 (white)		
	70/20 (red)		

a Hygienic effect when benzene was used as a solvent.

Four hundred possible referents remained after the age-matching. Among these, 40 persons with highly qualified work operations and 40 persons with less qualified work operations were randomly selected as referents.

Comments. In the choice of the first reference group a minimum of criteria of selection was used to attain a satisfactory variation of the background factors. However, an unintentional selection did occur in the first phase of work. The enterprise did not allow some groups of workers to be included in the reference group. The significance of this deficiency has not been investigated. Furthermore, it was later shown that many of the referents from departments with highly qualified types of work did not actually have very exacting work phases (storage work, cleaning work, etc). Obviously the collective assessment concerning education and occupation did not turn out satisfactorily because these departments had a more heterogeneous personnel than was expected. Therefore, the selected reference group was probably more heterogeneous than the exposed group with respect to different background factors.

During the project, assessments of the psychiatric health status and psychological performance level were carried out and showed alarmingly high frequencies of disorders that caused even more apprehension about unintentional mechanisms of selection. This suspicion was strengthened when a dropout of 39 % occurred in the primary selection of the reference group.

It was decided to obtain a description of the dropouts through an interview with the 51 persons who had refused to participate after being primarily selected. The interviews, of 15—20 min, took place at the workplace, 21 of the persons were not available for interview for different reasons — however only one actually refused to participate. Therefore, only 30 interviews were carried out. A comparison was made between the 30 interviewed dropouts and the reference group included in the study (reference group 1). Some differences (p < 0.05) were found that were important with respect to the suitability of reference group 1 as a reference group. Concerning medical history, a greater part of this group had had head injuries (15 %

to 0 %), other prolonged illness (15 % to 0 %), and illness which required surgical operation (46 % to 3 %). In addition a greater part of reference group 1 had been employed by the enterprise for less than 5 a (32 % to 7 %). The results revealed that this group had shorter employment periods in the enterprise than the dropouts and that more persons in reference group 1 had had short employment in different enterprises. Also, a greater part of reference group 1 (62 %) than the dropouts (23 %) had admitted using alcohol daily or once a week. Finally, a much larger proportion of reference group 1 (50 %) than the dropouts (26 %) was unmarried.

The summary assessment of the magnitude and structure of the dropout confirmed the suspicion of negative selection. Since the original aim of the selection did not seem to be fulfilled, a new reference group was chosen and examined.

Reference group 2

Methods of selection. At two other electronics plants in Stockholm, all the personnel (N = 800) was placed at the disposal of the project. The first step was to select the subjects with the same methods used for the first reference group. The second step was to make individual assessments concerning occupational classification, education, and employment stability based upon staff lists and inquiries.

So that the type of problems connected with the first reference group would be avoided, an individual matching with the exposed group was carried out. The criteria of this individual matching were (i) age (± 1 a), (ii) same educational level, (iii) position (foreman, not foreman), and (iv) employment stability (time of employment at the actual or previous workplace).

For each exposed person, lists of all the possible persons to be included in a reference group were drawn up. Among these potential referents, one person was randomly selected to be included in reference group 2 and one was chosen as a reserve in case the person first selected should refuse to participate for some reason in the investigation. Fifteen reserves were called to the examination. The reasons for the dropout of 15 of the primarily chosen persons are shown in table 3.

Comments. The introduction of individual matching limited the selection to the reference group. Thus there was a risk that the variation in the potential background factors tended to be too small. However, the risk was compensated for by the big pool from which reference group 2 was taken. The total dropout from reference group 2 was of the same order of magnitude as from the exposed group. In light of the fact that motivation to participate in an investigation is normally much lower in a reference group than in an exposed one, the results can be considered good. On the basis of the magnitude of and reasons for the dropout, it was considered that reference group 2 could be used for comparison with the exposed group. However, for the sake of completeness in the presentation of the results — despite the deficiencies — reference group 1 has also been included.

SCHEDULE OF THE MEDICAL EXAMINATION AND PRESENTATION OF THE RESULTS

Each of the selected subjects was examined at the Karolinska Hospital during two consecutive days following the schedule shown in table 4. Two subjects belonging to the same age stratum were examined on the same day, one from the exposed group and one from reference group 1. Through a period of two weeks 12 subjects were examined: 3 car painters, 3 industrial painters, 3 referents with highly qualified work and 3 referents with low qualified work. The examination was performed on weekdays (Tuesday—Friday).

For the reasons already presented, reference group 2 was examined separately, after the examination of the other two groups had been completed; however the same schedule was followed for this group also.

In the following text the results have been presented in the form of Z diagrams, which are the results of a statistical analysis of the differences between the exposed group and each of the reference groups.

The Z values have been calculated as follows:

$$Z = \pm (\overline{X}_e - \overline{X}_r)/S_p,$$

where $\overline{X}_e$ is the mean value of the variable in the exposed group, $\overline{X}_r$ is the mean value of the same variable in a reference group, and S_p is the standard deviation pooled within groups and age strata. For those variables in which the results were available as number of "yes" responses to a question on the questionnaires, the Z value was calculated as follows:

$$Z = (P_e - P_r)/\sqrt{P (1 - P) (1/N_e + 1/N_r)},$$

where P_e is the percentage of "yes" responses in the exposed group, P_r is the percentage of "yes" responses in a reference group, P is the total percentage of "yes" responses in both these groups, and N_e and N_r are the corresponding sample sizes.

The Z values can easily be transformed into a measure of significance of a more familiar type, as p-values ($p < 0.05$, < 0.01 and < 0.001).

The Z diagrams are presented in the form of bars showing the Z values for the different variables presented in the diagram and thus giving the difference (or more precisely the reliability of the dif-

Table 3. Reason for dropout from reference group 2.

Reason	Number
Prolonged illness	2
Previous exposure	1
Refusal	12
(among whom an expressed fear for hospitals and physicians)	(5)
Total	15

Table 4. Schedule for the different examinations.

Examination	Time required (h)
First day (0800—1730)	
Blood sample	0.5
Medical inquiry	1.5
Neurophysiological examination	2
Neurological examination	2
Psychological examination	2
Neuroradiological examination	0.5
Second day (0830—1630)	
Sociopsychological interview	1.5
Psychological examination	3
Psychiatric examination	1.5
Ophthalmologic examination	0.5
Total	15

ference) between the groups. A Z bar directed to the right in a figure means that the difference was regarded as indicative of a poorer state of health or level of function in the exposed group than among the referents, while a bar directed to the left means a better state in the exposed group.

GENERAL MEDICAL EXAMINATION

The general medical examination included a questionnaire concerning present and previous state of health (table 5) and a

Table 5. Content of the medical questionnaire.

Contents	Number of questions
Previous state of health	40
Present state of health	
General symptoms	23
Neurological symptoms	38
Psychiatric symptoms	52
Symptom present only at or immediately after work	11

number of blood sample analyses. As shown in table 5, the inquiry comprised questions on neurological and psychiatric symptoms. It must be stressed that the completed questionnaire was used as a basis for the interview at the following medical examination, during which the subject could clarify the answers given in the questionnaire.

Previous state of health

The Z diagrams in fig 2 illustrate the difference between the exposed and reference groups with respect to previous illness. It appears that the morbidity, in general, was somewhat higher in the exposed group as compared to the referents. Statistically significant differences were found for psychiatric symptoms ($p < 0.01$), as well as for infections of the urinary tract, gastritis and backache ($p < 0.05$), when the exposed group and reference group 2 were compared. The differences versus reference group 1 were somewhat smaller and sta-

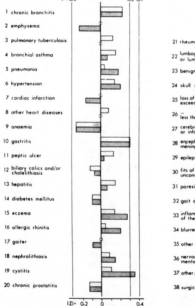

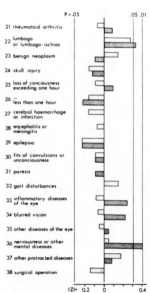

Fig 2. Differences, between the frequencies of "yes" responses in the exposed group and the two reference groups, concerning questions in the medical questionnaire regarding previous state of health. The differences are given as Z values. Light bars refer to reference group 1, and dark bars to reference group 2.

tistically significant only for gastritis (p < 0.05).

Present state of health

General symptoms

One group of 23 questions on the part of the questionnaire dealing with present state of health covered symptoms concerning the respiratory, circulatory, and digestive systems and the skin. The average number of "yes" responses was larger among the exposed than the reference groups. This difference was statistically significant with respect to skin trouble when the exposed group and reference group 1 were compared. When the frequency of "yes" responses in the groups were compared for separate questions, 7 of the 23 questions separated the exposed group from one or both of the reference groups. These particular questions and the differences are displayed in fig 3.

Another group of 11 questions concerned symptoms only present at or immediately after work. The following five acute symptoms were found more often among the exposed: vertigo, headache, difficulty in breathing, slurred speech, and feeling of intoxication (fig 4).

Psychiatric symptoms

In the analyses of the answers in the psychiatric part of the self-administered questionnaire, a statistically significant higher frequency of "yes" responses was found for the exposed group in comparison to the

referents for 8 of the 52 questions. They concerned increased fatigability, memory disturbances, absentmindedness, dysphoria, and nightmares (fig 5).

Neurological symptoms

The 38 questions in the neurological part of the inquiry dealt with symptoms from the central nervous system (11 questions), the cranial nerves (8 questions), and the peripheral nerves (19 questions). For this part of the questionnaire also, the average frequency of symptoms was higher for the

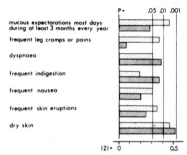

Fig 3. Differences, between the frequencies of "yes" responses in the exposed group and the two reference groups, concerning questions in the medical questionnaire regarding present state of health of the respiratory, circulatory, and digestive systems and the skin. The differences are given as Z values. Light bars refer to reference group 1, and dark bars to reference group 2.

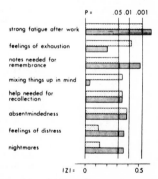

Fig 5. Differences, between the frequencies of "yes" responses in the exposed group and the two reference groups, concerning questions in the medical questionnaire regarding psychiatric symptoms related to present state of health. The differences are given as Z values. Light bars refer to reference group 1, and dark bars to reference group 2.

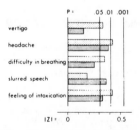

Fig 4. Differences, between the frequencies of "yes" responses in the exposed group and the two reference groups, concerning questions in the medical questionnaire regarding symptoms only present at or immediately after work. The differences are given as Z values. Light bars refer to reference group 1, and dark bars to reference group 2.

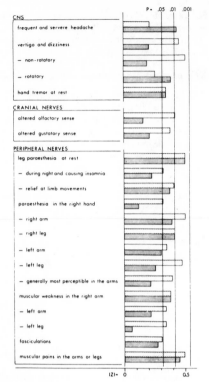

Fig 6. Differences, between the frequencies of "yes" responses in the exposed group and the two reference groups, concerning questions in the medical questionnaire regarding neurological symptoms related to present state of health. The differences are given as Z values. Light bars refer to reference group 1, and dark bars to reference group 2. (CNS = central nervous system).

cytes, leucocytes, and liver function tests [alcaline phosphatases, alanine aminotransferase (ALAT), asparate aminotransferase (ASAT) and bilirubin in serum]. The group mean values were all within current normal limits. However, some statistically significant differences were found between the exposed group and reference group 2; higher values were found in the exposed group for alcaline phosphatases (3.14 versus 2.72 μkat/l, p < 0.01) and for hemoglobin, hematocrit, and erythrocytes (p < 0.05).

Comments

It appears from the results that the solvent-exposed workers had a higher frequency of neurological and psychiatric symptoms than their referents. Also skin symptoms occurred more frequently among the exposed, a finding not surprising since it is well known that occupational exposure to solvents may cause irritative changes of the skin (14). Other differences in the previous and present state of health between the groups were spurious, and some of them (eg, some gastrointestinal symptoms) may merely have been a reflection of disturbances of neuropsychiatric health. The increase, although not statistically significant, in chronic bronchitis and some respiratory symptoms among the exposed may on the other hand be a manifestation of a solvent effect on the lungs, as recently demonstrated in epidemiologic (21, 58) and experimental studies (72).

exposed group than for the two reference groups. For 21 of the questions the difference was statistically significant with the exposed group versus one or both of the reference groups. These 21 questions are presented in fig 6, where it is demonstrated that the differences were the most pronounced for the questions concerning symptoms indicative of peripheral neuropathy (paresthesia and muscle pain).

Blood analyses

The blood analyses comprised sedimentation rate, hemoglobin, hematocrit, erythro-

PSYCHIATRIC INVESTIGATION

Acute psychoorganic syndromes are characterized by disturbed consciousness and hallucinations, whereas the chronic syndromes are distinguished by reduced mental performance, personality changes, and subjective symptoms. They may be individual reactions to reduced physical and mental ability and can be prominent in cases of a slight to moderate cerebral lesion or in the early phases of the psychoorganic syndrome. Moreover, subjective symptoms can occur much earlier than objective signs (54). The clinical picture is characterized by, eg, fatigue, worrying, and memory and concentration difficulties, and it is often described as a neurasthenic syndrome. Similarly, subjective symptoms with a concurrent and varying reduction in mental capacity following long-term exposure to organic solvents have been reported in a great number of case reports and in epidemiologic studies (59). Therefore, the present study included a psychiatric assessment of a number of neurasthenic symptoms. However, the neurasthenic syndrome has been attributed to psychological and social factors. Furthermore, such factors can be influenced by organically induced personality changes (83). Hence, the present study included investigations of certain personality traits and of the adjustment of the individual to social life.

Methods

Questionnaires

The self-administered psychiatric questionnaires contained the 52 questions described in the section on the general medical examination and 29 questions about the social situation of the individual according to Gardell (30) and Westlander (85). In addition, five questions concerned the use of alcohol. The information so obtained was used as the basis for the psychosocial and psychiatric interviews. Certain questionnaires about the individual's habitual behavior (inventories) were also presented so that some personality traits could be studied which had been postulated by factor analyses [Eysneck's Personality Inventory (EPI)] or were based on theoretical considerations (Marke-Nyman's Temperament Scale (MNT)].

Psychosocial interview

The structured interview contained 20 items which, together with the aforementioned questions, have been used in previous studies (30, 85). Questions and items of similar content were brought together in 18 indices. These 18 indices were assembled into three general indices giving an overall assessment of the conditions at home, those at work, and any signs of social maladjustment (eg, abuse, criminality).

Psychiatric interview

The prevalence of various symptoms experienced during the last six months was rated according to the examinee's reports given during a semistructured interview (with a duration of approximately 45 min). Thirty-three reported items were rated on a seven-step scale. Twenty-one of the items were drawn from the CPRS (Comprehensive Psychopathological Rating Scale) (8). The 12 remaining items were constructed according to the same principle. Also, the interviewer rated his own observations of 12 symptoms and signs observed during the interview. He assessed the "assumed reliability" or "credibility" of the reports given during the interview and, finally, the overall mental health of the subject (two items). The psychiatric interviews were recorded on tape to permit assessment of observer bias and interrater reliability. Individuals in the exposed group and the first reference group were interviewed in random order by two senior psychiatrists. Individuals in the second reference group were also interviewed by a resident in psychiatry. Each one of the three interviewers of the second reference group rated approximately one-third of all the individuals in every age stratum. As regards the exposed group and reference group 1, the interviewers endeavored to remain blind as to the nature of the examinee's work. However, the second reference group was examined separately and the individual work was known by the interviewers. In all cases, any information about type of work was erased from the tapes and could therefore not influence later reratings. Thirty-five tapes were selected at random and rerated (five per age stratum and rater). This rerating took place 1 to 2 a after the original interview. The product-moment correlation (r) between the raters was computed from the sum score of all the reported items. It varied between 0.82 and 0.88. The correlation (r) between the original ratings and reratings performed by the same interviewer varied between 0.92 and 0.95. The average score of the rerating was very close to that obtained at the original rating. The two senior psychiatrists arrived at almost exactly the same score when rerating the same tape. The resident, however, gave higher scores. The interview technique used may be considered to be "conspective" (ie, it gives the same result regardless of the rater, at least between senior psychiatrists). In the statistical procedures the original ratings were used, with some exceptions. Multivariate statistics were applied to 13 selected items drawn from the ratings made by the senior psychiatrists. This material consisted of

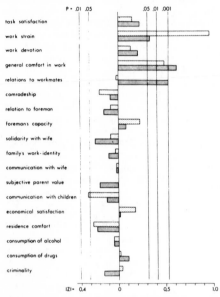

Fig 7. Differences between the mean values of the psychosocial indices of the exposed group and the two reference groups. The differences are given as Z values. Light bars refer to reference group 1, and dark bars to reference group 2.

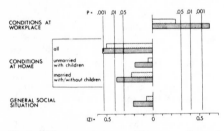

Fig 8. Differences between the mean values of the general psychosocial indices of the exposed group and the two reference groups. The differences are given as Z values. Light bars refer to reference group 1, and dark bars to reference group 2.

ratings from 77 exposed and 127 reference workers.

Results

Questionnaires

The frequencies of certain mental complaints as expressed in the medical questionnaire have already been given. As for the personality inventories (EPI & MNT), the results agreed with those of previous normalization studies (11, 55). Neuroticism in the EPI and validity in the MNT were of particular interest, since they can be regarded as reflecting a constitutional disposition for the development of neurotic reactions. However, neither the exposed nor the two reference groups seemed to be handicapped by any such disposition. The results of the inventories speak against selective factors (eg, choice of work) influencing the psychological comparativeness of the groups. If such factors had been operating, they would have induced difficulties in the assessment of the psychiatric and the psychosocial interviews. (We have made an exception, however, of the factors operating in the composition of the first reference group).

Psychosocial interview

A considerably higher work stress was reported by the exposed group in comparison with both reference groups (fig 7). When car painters and industrial painters are compared only with their subgroup among reference individuals, this difference persists. The exposed individuals reported a greater subjective work stress than did the reference groups; this finding was attributed to experienced risks at work (not only those connected with organic solvents). The exposed workers clearly reported a lower general ease in their work conditions. They also experienced more difficulties than reference group 2 in their relationships with fellow workers. The remaining indices revealed no differences. As expected, the exposed group reported more complaints about their work situation than the referents when general indices were compared (fig 8). In a comparison of the exposed and the first reference group a trend was observed which became a statistically significant difference when the same comparison was made with the second reference group. However, the exposed individuals reported more favorable conditions at home than both reference groups. This finding was particularly marked for the married painters when they were compared to married individuals in reference group 2. The general index for disturbed adjustment to social life, which contained criminality, alcohol and drug abuse, did not differ between the groups.

Psychiatric interviews

Reported items. Fig 9 shows the results of the ratings of the 33 reported items. In comparison with both reference groups, the exposed group had a clearly higher symptom score for 11 items (inner tension, irritability, fatigability, aches and pains, learning problems, short- and long-term memory problems, nausea, epigastric pain, headache and, finally, precision movement). Statistically significant differences between the exposed group and one of the reference groups were found for five additional items (worrying over trifles, concentration difficulties, vertigo and dizziness, dyspnea, and mood lability). The most pronounced differences were observed for memory problems, headache, and fatigability. The figure shows, too, that there were trends towards differences for several other items between the exposed and reference groups.

For further statistical analysis, a reduction of the amount of data was desired. Those 13 items were selected that were clearly correlated to items that best differentiated between the groups ($r > 0.3$, $p < 0.001$). These 13 items correlated clearly with group residence "exposed" (multiple $r \sim 0.54$, $p < 0.001$).

The correlation was mainly derived from symptoms such as short-term memory problems, headache, mood lability, and precision movement difficulties (table 6). Furthermore, the items were subjected to factor analysis. In the search for symptom clusters with a common variance related to exposure to organic solvents, the analysis was confined to ratings obtained in interviews with the painters. According to the factor structure, 13 items were grouped into three general indices (fig 10). The first index showed the most prominent dif-

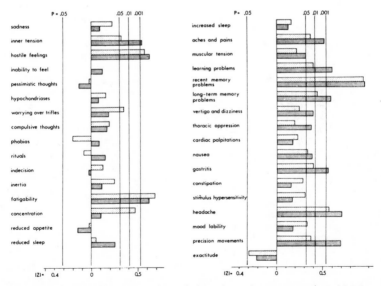

Fig 9. Differences between the mean values of the ratings of symptoms reported by the exposed group and the two reference groups. The differences are given as Z values. Light bars refer to reference group 1, and dark bars to reference group 2.

238

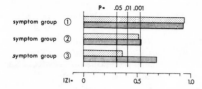

P= .05 .01 .001

symptom group ①
symptom group ②
symptom group ③

IZI= 0   0.5   1.0

Fig 10. Differences between the mean values of the three psychiatric general indices. The differences are given as Z values. Light bars refer to reference group 1, and dark bars to reference group 2. Symptom group 1 comprises fatigue, concentration difficulties, memory problems, vertigo, gastritis, and headache. Symptom group 2 comprises hostile feelings, worrying over trifles, indecision, inertia, thoracic oppression, and inner tension. Symptom group 3 consists only of precision movement difficulties.

Table 6. Multiple regression analysis of selected psychiatric items.

Step	Item	Multiple r	p
1	Short-term memory problems	0.454	0.000
2	Headache	0.493	0.002
3	Mood lability	0.512	0.028
4	Precision movement difficulties	0.522	0.105

ference between the groups and contained the largest factor, corresponding to 25 % of the total variance.

Observed items. Generaly, the observed items displayed less difference than the reported items between the examined groups. The following differences were obtained: The word fluency of the painters was significantly higher, and they displayed problems with their short-term memory significantly more often than the reference groups. In comparison with either reference group, there were also signs of long-term memory problems, increased motor restlessness, and increased muscle tension in the exposed group. The general credibility of the ratings was scored as high for all groups.

Relationship with psychosocial findings. Some differences between the exposed and the reference groups were observed for certain psychosocial items, as well as certain mental complaints. A relationship between psychosocial and psychiatric findings is essential for the interpretation of the origin of the psychiatric symptoms.

The existence of neurotoxic agents such as organic solvents in an individual work environment can give rise to mental symptoms. Such symptoms can lead to increased subjective work stress and social adaptation problems. On the other hand, high work stress and unfavorable conditions at work can also provoke mental symptoms such as fatigability. However, the relationships are probably very complicated, and only a rough assessment can be attempted in this report.

An examination of the relationship between the general psychosocial and psychiatric indices of the two reference groups can provide an estimate of the extent to which the psychiatric ratings were influenced by the examinee's experience of his environment at work and at home. No such correlations were noted. However, the general index for work situation contained one index for work strain (fig 7), which showed a weak correlation with the first symptom group ($r = 0.21$ $p < 0.006$). No correlations between the psychiatric rating and the remaining separate psychosocial indices were found. Therefore it appears that the ratings of the psychiatric symptoms were influenced only to a small extent by the examinee's experience of, eg, work conditions.

As for the exposed group, only a slightly higher correlation between work stress and psychiatric symptoms was found ($r = 0.35$ $p < 0.36$). The index for subjective work stress contains items relevant to experienced risks at work. Of course the painters may be more concerned over work risks, and this concern can explain the difference between the two groups.

Discussion

The psychiatric investigation showed that psychiatric complaints were considerably more common among painters than among the individuals in both reference groups. This result could be due to a tendency to give exposed subjects higher scores. This contention, however, is contradicted by the findings of high rating-rerating reliabilities and of the observation that the interviewing psychiatrists arrived at approximately the same average scores. Some individuals in reference group 2 were interviewed by a psychiatric resident, who gave higher average scores than the two specialists. This occurrence induced a systematic error which, however, would suppress observed differences. Reference group 2 has thus been given ratings that are somewhat too high.

At the interviews, the painters did not give impressions of overstating their complaints. The presence of such a tendency would have influenced the general credibility of the interviews. However, in this respect, painters and reference workers were equal. The possibility remains, however, that the selection of the groups can have induced considerable differences between the painters and reference workers with respect to certain background factors that could affect their mental state. The inventories used in the study did not reveal personality traits likely to cause mental symptoms. The psychosocial investigation showed that the painters viewed their work situation more negatively and experienced their work situation as more negative than did the referents. However, since the correlations were weak, they cannot explain the higher incidence of mental complaints among the exposed workers. Therefore, a relationship between mental symptoms and exposure to organic solvents seems to be a more plausible presumption. This view is supported by the observations that those symptoms that best differentiated between the groups are also common in persons with slight cerebral damage (24). The symptomatology also agrees with previous case reports or epidemiologic studies (59). A closer analysis of the ratings indicates that memory difficulties, the single item that best differentiated between the groups, was connected with a number of cerebrolesional symptoms (83).

The CPRS (1) formed the basis for the used psychiatric rating scale. The scale should be of the interval type (79) to permit the proper use of parametric statistics, but any imperfection in this respect would have little bearing on the results (6, 19). Similarly, certain theoretical objections can be raised against the use of factor analysis of ratings, particularly if the object is to give the factor structure a mathematical interpretation. However, the actual use of the method for data reduction is generally accepted (16, Schalling: personal communication).

The stability of the factor structure must be regarded with caution. The comprehensive psychiatric indices were based on factors of first order, and it may be difficult to reproduce them exactly with other materials.

Furthermore, the validity of the rating scale has not been tested. It is of interest to note that preliminary data show significant relationships between the first psychiatric index and results of psychometric tests indicative of cerebral damage. The third psychiatric index also showed a statistically significant correlation with a psychometric test demanding manual dexterity. These correlations will be the subject of further analysis.

PSYCHOLOGICAL EXAMINATION

A fairly large number of studies has shown that experimental exposure to solvent vapors can cause changes in central nervous functions as measured by behavioral performance tests. Thus, for several solvents, performance decrement on tests of, eg, sensory, perceptual and psychomotor functions was found to be related to the concentration of solvent in inspiratory air and to the total uptake of solvent in the organism (2, 27, 28). The results of these and other experimental investigations have provided a basis for the evaluation of the risks for acute effects due to the presence of solvents in the work environment.

The risk for decrement in performance capability due to long-term exposure has not yet been satisfactorily evaluated. Epidemiologic investigations of occupationally exposed workers do, however, clearly indicate that long-term exposure to solvent vapors can cause a decrement in behavioral performance capability (29, 33, 37, 38, 39, 44, 52, 53).

Methods

A total of 18 different behavioral performance tests were used in this investigation (table 7). Fourteen of these tests were from a recently standardized test battery for investigations in behavioral toxicology (80), and four were tests on electronic apparatus which were developed in the work psychology divison of the Swedish National Board of Occupational Safety and Health (44).

Except for the tests on the electronic apparatus, the factor structure of the test battery was already known from previous investigations (80). A number of factor analyses was however performed also on the results from the present investigation (25). These analyses showed that the factor structure for the exposed and reference groups was the same as for the population of an earlier investigation (80). In these latter analyses, all of the tests from the present investigation had been included, and thus an analysis of the relations between the tests of the battery and

the tests on the electronic apparatus was possible.

Statistical analysis

From the first inspection of the data, it was obvious that the results of all the tests except the two verbal ones were highly correlated with age. This fact, and the heterogeneity of the groups with respect to age, would make the t-testing of the mean differences between groups inadequate. A t-test would result in a highly underestimated significance level. The statistical analyses were therefore made with two-way analyses of variance with exposure and age as the two sources of variation. This method was preferred to alternative models of analysis, such as the t-test for matched pairs or an analysis of covariance. The analysis of variance was chosen because of the fact that it gives information about a possible effect of interaction between exposure and age. Furthermore, this model does not require

Table 7. Means and standard deviations from different tests for the reference and exposed groups and for the car and industrial spray painters.

Test	Exposed group (N = 80)		Reference group (N = 80)		Car painters (N = 40)		Industrial painters (N = 40)	
	Mean	SD	Mean	SD	Mean	SD	Mean	SD
Synonyms	19.8	6.1	20.5	5.8	19.7	6.4	20.0	5.9
Distinction	16.1	6.3	16.2	6.3	15.9	6.5	16.3	6.2
Spatial transposition	26.0	7.9	26.5	7.4	25.9	7.2	26.1	8.6
Block design	22.5	7.3	24.7	6.4	22.2	7.3	22.9	7.3
Figure classification	20.7	4.6	21.2	4.2	19.8	4.2	21.6	4.8
Visual gestalt ability	134	21	137	17	137	19	131	23
Digit symbol	39.5	13.2	44.9	11.9	39.5	12.7	39.5	13.7
Identical numbers	79.0	17.5	85.9	17.0	79.3	15.3	78.8	19.6
Dots	654	145	712	141	656	132	652	159
Mental arithmetic	2.6	1.3	2.3	1.0	2.5	1.2	2.7	1.3
Benton (reproduction)	6.9	1.5	7.3	1.5	6.7	1.4	7.0	1.7
Memory test (recognition)	28.1	4.1	28.9	4.1	28.7	3.7	27.6	4.3
Memory test (reproduction)	52.7	9.6	56.5	6.2	53.9	8.4	51.5	10.6
Hand dexterity test	76.8	8.0	80.7	8.8	77.8	7.0	75.7	8.8
Finger dexterity test 1	26.0	6.0	29.4	6.4	27.2	4.9	24.8	5.0
Finger dexterity test 2	26.4	5.0	29.0	5.8	27.2	4.9	25.7	5.0
Simple reaction time, speed	264	28	243	30	258	29	270	29
Simple reaction time, decrement	24	27	14	23	23	24	24	30
Finger tapping	203	26	202	28	209	30	197	21

any assumption about a linear relationship between performance and age.

This first inspection revealed that reference group 1 was much more heterogeneous in performance than reference group 2 or the exposed group. Reference group 1 thus showed a greater standard deviation than the other two groups on almost all performance tests. Furthermore the distributions of the performance scores for reference group 1 showed a marked negative skewness in most of the tests. These observations accentuated the interpretation of reference group 1 as being negatively selected. Because of these facts, the results of reference group 1 were excluded from the analysis of variance.

In the analyses of variance, the material was grouped into five age classes according to the sampling criteria. Four groups were used, car painters, industry painters and one matched reference group for each of the exposed groups. The resulting design was 5×4 with eight individuals in each cell. This grouping made possible the simultaneous testing of the differences between the two exposed groups and between each of these and the reference groups. Combining the two exposed groups and the two reference groups, respectively, made it possible also to make a test for interaction between exposure and age.

The reaction-time test, Simple RT, could also be analyzed for performance changes during the test period. For this test, a two-way analysis of variance with repeated measurements was employed in which the trend of the performance changes over time was introduced as a further source of variation.

For more information concerning the correlation between groups (exposure/no exposure) and performance on different tests, a series of multiple regression analyses was performed. These analyses resulted in a multiple correlation coefficient for the relationship between groups (exposure/no exposure) and performance variables from the same factor (table 8).

Results

The means and standard deviations for the different groups and performance tests are shown in table 7. The relative magnitude and direction of the difference in performance between the exposed group and the reference groups are illustrated in fig 11, where the mean differences for each test have been transformed into standardized scores.

The result of the analyses of variance showed that the main sources of variation,

Table 8. Multiple correlation coefficients for the relation between performance on tests belonging to the same factor and group membership (exposed group/reference group).

Factor	Multiple r	p
Verbal comprehension	0.070	0.700
Reasoning and spatial relations	0.187	0.158
Perceptual speed	0.230	0.047
Numerical ability	0.147 [a]	0.072
Memory	0.240	0.034
Manual dexterity	0.281	0.007
Simple reaction time	0.360	< 0.001
Motor speed	0.031 [a]	0.703
All the tests	0.529	< 0.001

[a] Pearson correlation coefficients for factors represented by only one test.

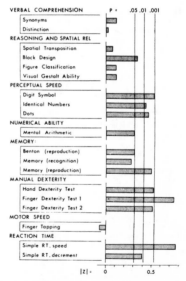

Fig 11. Mean performance differences between the exposed group and reference group 2 on the different psychological tests. The differences are expressed in standardized scores. (REL = relations, RT = reaction time)

exposure and age, had a significant influence on performance on most tests. On the other hand, there were no signs of any interaction between exposure and age. The performance of the older subjects was worse on the tests irrespective of exposure.

With one single exception the mean performance on the tests was poorer for the exposed group than for the reference group (fig 10). From fig 10 it can be seen that the exposed group had a significantly

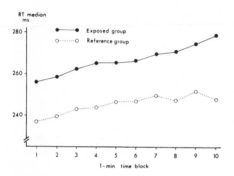

Fig 12. Change in mean reaction time (RT) over time (1-min time blocks) for the exposed group and reference group 2. Each point is a mean of the individual median values for the time block.

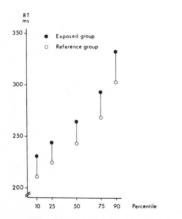

Fig 13. Mean values of the reaction time (RT) of the exposed group and reference group 2 at different percentiles in the individual distributions of reaction times.

poorer performance than the reference group on all three tests of perceptual speed (p < 0.01) and manual dexterity (p < 0.01). The test of Simple RT also showed a distinct difference between the two groups (p < 0.001).

The results of the memory tests also indicated poorer performance for the exposed group as compared to the reference group. The mean difference in performance was clearly significant (p = 0.003) for one of the tests (memory, reproduction) and very close to significance (p = 0.057) for another test (Benton, reproduction).

The group differences in cognitive functions such as reasoning, spatial relations, and numerical ability were more ambiguous. The four tests of reasoning and spatial relations showed only one significant difference (block design test, p = 0.046), while the differences for the others were small. A tendency towards a performance difference can be seen in the results of the test of numerical ability (mental arithmetic, p = 0.070). Tests of verbal comprehension and motor speed did not differentiate between the groups.

From table 7 it can be seen that both car painters and industrial painters generally performed less well than the reference group. However, there was a clear tendency for the industry painters to perform still worse than the car painters on the tests for manual dexterity and simple reaction time. Manual dexterity and reaction time are also the functions for which the greatest and most unambiguous differences between the exposed groups and the reference group were found. This fact is illustrated in table 8, where the correlation between exposure and performance is described with reference to the factor structure of the test battery. It can be seen that reaction time, manual dexterity, memory, and perceptual speed clearly correlated with exposure. In addition, the simple reaction time test obviously had nearly half of the discriminative power of the test battery (0.360^2 as compared to 0.529^2). Fig 12 and 13 give a more detailed description of the results for the reaction time test. Fig 12 illustrates the change in reaction time over the test time (10 min) for the exposed and reference groups. As already mentioned, the difference in reaction time between the two groups was significant (p < 0.001). Furthermore, the analyses of variance and trend showed that the change in reaction time over time was approximately linear (p < 0.001), and the change was significantly greater (p = 0.024) for the exposed group. These analyses also pointed out that still another aspect of reaction-time performance correlated with exposure. This phenomenon is illustrated in fig 13, in which the mean values of each group have been presented for the different percentiles in the distribution of reaction times of every individual in the test. As can be seen in the figure, the mean differences between the groups increased over the percentiles (p = 0.025). Therefore the individual distributions of reaction time tended to be more positively skewed in the exposed group than in the reference group.

Discussion

Unambiguous differences in performance between the exposed and reference groups were found in the psychological examinations. The exposed group performed significantly less well on tests of functions such as simple reaction time, manual dexterity, perceptual speed, and memory, ie, in not very complex perceptual and psychomotor functions.

The results in general showed a clear relationship between exposure and performance on the tests. This correlation also appears to be more unambiguous in this investigation than in the earlier studies of house painters (39) and car painters (53). It is not probable that the groups differed initially with respect to intellectual ability because no differences between the groups were found in the more intellectual tests of the battery, eg, verbal comprehension. The results also seem to favor the hypothesis that verbal ability is relatively resistent to the effect of solvent exposure (39). In a previous study of car painters (53), differences between an exposed and reference group were found for all the functions tested but reaction time, while in this investigation the greatest difference between the exposed and reference groups was found on the reaction-time test. This is a very im-

portant difference because it is well known that performance on a reaction-time test is independent of intellectual capacity. This is not the case for the tests of other functions. Thus the result on the test of reaction time can be interpreted as due to solvent exposure, even if the comparability of the groups in intellectual ability is questioned.

As already mentioned, the exposed group performed less well also on the memory tasks. It should be stressed, however, that performance on these and other available memory tests is highly dependent on the attention and concentration of the subject. Furthermore, retention was measured after a very short time. The poorer performance on these tests is therefore not necessarily related to the subjective memory impairment reported by some exposed individuals.

The effects of long-term exposure to solvents already described seems to be of the same kind as those found in laboratory investigations of acute exposure (27).

The analyses of variance illustrated a negative correlation between age and performance on the test. However, there was no effect of interaction between age and exposure. Thus the differences in performance between the exposed and reference groups were relatively stable with respect to age. The variation of age in the exposed group was almost identical with that of length of exposure (painter years). Therefore it was not possible to demonstrate any effect or correlation that indicated a dose-response relationship. Nor has such a relationship been found in the earlier investigations of house painters (39) and car painters (53).

The general results can, on the basis of a cursory analysis, seem somewhat contradictory. While on one hand unambiguous mean differences occur between exposed and reference groups, there are on the other no correlations between length of exposure and test performance. However, there is the possibility that the exposed population has been subject to a systematic dropout, a "healthy worker effect." It is obvious that it would be very difficult to prove a dose-response relationship in such a selected population. These results would also be explained if the effects are subchronic rather than chronic.

NEUROLOGICAL AND NEUROPHYSIO-LOGICAL EXAMINATIONS

There are several reports in the literature on objectively demonstrable lesions in the central nervous system due to occupational exposure to organic solvents. Even recently, relatively serious functional disturbances of the central and peripheral nervous systems have been reported due to such exposure (3, 15, 18). The exposure conditions described in these studies seem to have been worse than in comparable Swedish industries. However, the results of recent Scandinavian epidemiologic studies (44, 45, 66, 75, 77) indicate that even a low degree of long-term exposure to industrial solvents can affect the nervous system to the point that changes can be detected in neurological and neurophysiological examinations. The changes

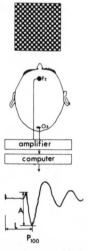

Fig 14. Visual evoked responses (VER) — At the top of the figure is the square checkerboard pattern screen with which the "pattern reversal" stimulation was performed. Below it is the skull of the examined subject (seen from above) with the two recording electrodes. Fz and Oz indicate the electrode positions according to international electroencephalographic standards. The recorded signals pass through an amplifier and a computer to give a response (VER) as shown at the bottom of the figure. Two variables were measured from this response, the amplitude (A) and the latency (L). P_{100} is the dominating positive wave since its latency is around 100 ms in normal cases.

found with these methods have been slight, eg, an increased occurrence of slight, peripheral neuropathy — mainly sensory — and nonspecific electroencephalographic (EEG) changes.

Methods

Neurological examination

In the clinical neurological examination, the function of the central nervous system (aphasia, gait, coordination, tremor, Romberg's sign, muscle tone, Babinski's sign) and the cranial nerves (sense of smell, eye movements, nystagmus, facial movements and sensibility) was tested. The peripheral nervous system was assessed with an examination especially designed to detect early signs indicative of polyneuropathy (12, 44, 45).

Sensibility (pain, temperature, discriminative sensibility and kinesthetic sense), tendon reflexes, and muscle strength were evaluated. The findings were scored as 0 = normal, 1 = mild changes, and 2 or 3 = varying severity of manifest disease.

Electroencephalography

The EEG recording was performed according to international standards; it included hyperventilation for 3 min and photic stimulation. Drowsiness was prevented by sound stimulation. For the computer analysis, the EEG signals were stored on magnetic tape.

The EEGs were visually examined by two neurophysiologists independently and without knowledge of whether the EEG belonged to the exposed or the reference groups.

A representative 10-s section of each EEG was selected and ranked with regard to the amount of low-frequency activity. The stored EEG signals were computer-processed by means of spectral parameter analysis (SPA) (84). SPA describes the frequency content of the EEG signal in one to three components (the delta, alpha and beta components). Each component is characterized by three parameters (peak frequency, bandwidth and power).

Visual evoked responses

Pattern reversal in visual evoked responses (VER) was recorded. The stimulus was produced by a commercially available device (Digitimer D 110); it consisted of a back projection of a black and white checkerboard pattern with a reversal frequency of 2 Hz (fig 14). [For details of the stimulus parameters see Nilsson (63)]. A bipolar recording of VER was made with two scalp electrodes, one applied at the midline 5 cm above the inion and the other placed midfrontally. Responses to 200 or 400 reversals were averaged. The pattern reversal VER has a typical waveform and good intraindividual reproducibility. The interindividual variability of the latency is small. In the analysis of the responses, the latency and amplitude were measured as is shown in fig 14.

Electroneurography

Six peripheral nerves (median, peroneal and sural nerves on both sides) were examined electroneurographically. Conduction velocities of motor nerves, as well as conduction velocities and action potential amplitudes in sensory and mixed nerves, were measured with surface electrodes (table 10). The electrode positions are presented in fig 15.

Vibratory perception thresholds

The vibration thresholds were determined at three different locations of the extremities, ie, the bony parts of the dorsum of the foot (tarsal), the lower leg (tibial), and the wrist (carpal). An electromagnetic biothesiometer with a plastic stimulator shaft, 6 mm in diameter, provided a 100-Hz sine wave stimulus movement, the amplitude of which could be varied between 0 and 25 μ. An accelerometer mounted on the stimulator shaft registered peak to peak amplitude of the stimulation, which was monitored digitally during the procedure.

Results

Neurological examination

Significant differences between the exposed group and reference group 2, with more positive findings in the exposed group, were found for the following: Romberg's test, finger-to-nose test, finger-to-finger test, and finger tremor. In all cases the findings were minor, and, since the examinations were performed by two physicians, for whom the distribution of exposed and reference subjects was unequal, the results must be regarded with caution.

The quantified neurological status concerning the function of the peripheral nerves did not reveal any significant differences between the groups.

Electroencephalography

As shown in table 9, the visual evaluation of the EEGs did not reveal any significant differences between the exposed group and the referents. The proportion of abnormal

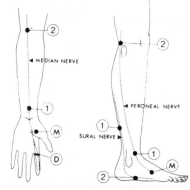

Fig 15. Electroneurography — The figure shows the position of the electrodes in the electroneurographic examinations. (M = muscle response recording electrode, D = digital stimulating electrode)

Table 9. Visual evaluation of the electroencephalograms (EEG).

EEG evaluation	Exposed group (N = 80)	Reference group 1 (N = 80)	Reference group 2 (N = 80)
Normal	44	47	47
Not definitely abnormal	33	23	28
Slightly abnormal	2	2	3
Fairly slightly abnormal	0	3	1
Moderately abnormal	1	5	1

246

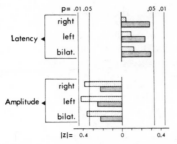

Fig 16. Differences between the mean values of the visual evoked response variables of the exposed group and the two reference groups. The differences are given as Z values. Light bars refer to reference group 1, and dark bars to reference group 2. (bilat. = bilateral)

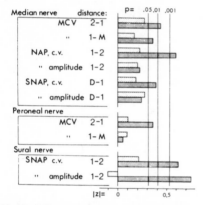

Fig 17. Differences between the mean values of the electroneurographic variables of the exposed group and the two reference groups. The differences are given as Z values. Light bars refer to reference group 1, and dark bars to reference group 2. The notations of the distances along the nerve refer to the electrode positions in fig 15. MCV = motor nerve conduction velocity, NAP = nerve action potential (in mixed nerves), SNAP = sensory nerve action potential, c.v. = conduction velocity).

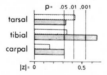

Fig 18. Differences between the mean values of the vibration thresholds of the exposed group and the two reference groups. The differences are given as Z values. Light bars refer to reference group 1, and dark bars to reference group 2.

EEGs was of a magnitude expected for a population of this kind. When the EEGs were ranked by one of us (AW), the distribution of exposed subjects and referents indicated that the former showed more alpha activity than the latter. This tendency was statistically significant ($p < 0.01$, Mann-Whitney rank correlation test). However, when a new ranking was made by two of us (HP and LW) jointly, no such difference was demonstrated.

The computer analysis (SPA) of the EEGs did not show any significant differences between the groups, but some tendencies were found which were in agreement with the results of AW's ranking: higher alpha power, smaller alpha bandwidth, lower delta power and larger delta bandwidth in the exposed group.

Visual evoked responses

Fig 16 shows the VER variables of the exposed group compared with those of the reference group. The latency was somewhat longer in the exposed group, especially in comparison with reference group 2, but the differences were not statistically significant. The mean amplitude of the exposed group was somewhat larger than that of the reference groups, particularly in comparison with reference group 1, for which the difference was statistically significant ($p < 0.05$). If the mean values of the latencies and amplitudes of the groups are divided into the five age strata, the difference in latency between the exposed subjects and the referents becomes present only in the older strata. The VER amplitudes did not show such an age dependence. However, a nonsignificant tendency towards larger differences was present in the youngest age stratum.

Electroneurography

The mean values and standard deviations for the electroneurographic (ENeG) variables are presented in table 10. Fig 17 shows the differences in these variables between the exposed group and the two reference groups. Seven of the 10 ENeG variables had significantly lower mean values for the exposed group in comparison with reference group 2. The differences in all the ENeG variables were the largest for the youngest age stratum.

Vibratory perception thresholds

As shown in fig 18, the exposed group had higher vibration thresholds than the reference groups. The difference was the most pronounced for the legs; it was similar in all age strata.

Discussion

Electroencephalography

No statistically significant differences between the exposed subjects and the referents were found in the visual evaluation of the EEGs. This result agrees with the findings of Seppäläinen et al (75) on a group of 102 car painters. However, other studies have revealed EEG differences between groups of subjects exposed to organic solvents and referents, but the type of EEG changes has been different in different studies. Knave et al (44) found a smaller amount of alpha activity; Rosén et al (67) found increased beta activity; Ekberg et al (20) found (with a computer-based frequency analysis) a somewhat lower alpha frequency; and Seppäläinen et al (76) found an increased amount of theta activity in the exposed group. A difference between the groups was also found in our material after a ranking of the EEGs, performed by one of us (AW), showing a greater amount of alpha activity in the exposed group. The EEG changes found in the aforementioned studies have been subtle and noncharacteristic, and the differences between exposed subjects and referents have appeared as statistical differences between groups.

No definite dose-response relationships have been found in the aforementioned works, but in an earlier study on styrene-exposed workers Seppäläinen & Härkönen (74) found abnormal EEGs — mainly an increased amount of slow activity — in

Table 10. Electroneurographic results (mean ± standard deviation).

Variable [a]	Exposed group	Reference group 2
Median nerve		
Motor nerve conduction velocity (MCV)		
Distance 2-1 (m/s)	54.29 ± 3.93	55.91 ± 4.09 **
Distance 1-M (factor) [b]	1.85 ± 0.31	1.94 ± 0.26 *
Mixed nerve action potential (NAP), distance 1—2		
Conduction velocity (m/s)	61.46 ± 4.28	63.81 ± 4.82 ***
Amplitude (μV)	26.74 ± 13.87	31.75 ± 15.81
Sensory nerve action potential (SNAP), distance D-1		
Conduction velocity (m/s)	53.13 ± 5.67	54.33 ± 5.79 *
Amplitude (μV)	15.90 ± 8.34	17.75 ± 9.06
Peroneal nerve		
Motor nerve conduction velocity (MCV)		
Distance 2-1 (m/s)	46.32 ± 3.91	47.96 ± 5.44 *
Distance 1-M (factor) [b]	1.43 ± 0.29	1.45 ± 0.20
Sural nerve		
Sensory nerve action potential (SNAP), distance 1—2		
Conduction velocity (m/s)	40.32 ± 4.19	43.10 ± 4.77 ***
Amplitude (μV)	7.23 ± 5.14	11.82 ± 7.95 ***

[a] The distances given in the table are defined by the two points delimiting the distance in question. These points were placed as shown in fig 15. All recordings were performed with surface skin electrodes and concerning NAPs and SNAPs bipolarly with both electrodes along the nerve and with an interelectrode distance of 20 mm.
[b] Factor = (distance/latency for muscle response) · 0.1 if distance is given in millimeters and latency in milliseconds.
* $p < 0.05$, ** $p < 0.01$, *** $p < 0.001$ between the two groups. It should be stressed that, for these significance calculations, standard deviations pooled within groups and age strata were used. The standard deviation values given in the table were not pooled.

one-third of the subjects with a urinary mandelic acid concentration of more than 700 mg/dm³, whereas the prevalence of abnormal EEGs among the workers with a lower level of mandelic acid was only about 10 %.

Visual evoked responses

Several different functions are tested by means of VER, ie, the transmission in the optic radiation, as well as synaptic transmissions in the retina, the lateral geniculate body, and the cerebral cortex. Abnormal delay of the responses is usually supposed to indicate a lesion of the optic nerve. In the exposed group, a tendency towards such a delay was found, though not on a statistically significant level. Organic solvents are not considered to produce the type of neuropathy (demyelinization) associated with a marked reduction of conduction velocity. Thus, a pronounced delay of the VER was hardly to be expected. The difference found concerning the amplitude of the VER (a larger mean amplitude in the exposed group than in the reference groups) is difficult to explain. Seppäläinen et al (76) found differences in flash-evoked VER between a group of n-hexane exposed workers and a reference group. Their results partially agree with ours, but the comparison is a difficult one due to differences in the stimulation technique used.

Electroneurography

The earliest subjective symptoms of neuropathy caused by long-term low level exposure to organic solvents are sensory; they appear — in agreement with the distribution of the pathological-anatomical changes — in the distal parts of the limbs, beginning in the feet (70, 73). In agreement we found the most significant (p < 0.001) differences between the exposed group and the referents in the function of the purely sensory sural nerve distally on the lower leg. However, we did not find any distal-proximal gradient for the peroneal nerve or the arm nerves. Unexpectedly, in the arms, the conduction velocity of the sensory nerves was less influenced (p < 0.05) than that of the motor fibers (p < 0.01) and the mixed nerves (p < 0.001).

In general, it is surprising that the conduction velocities were significantly affected while the amplitudes of the median nerve action potentials did not show any significant differences between the groups (fig 16). Previous experimental work has shown that most neurotoxic compounds cause axonal degeneration (8, 18, 70, 71).

In this type of nerve lesion, the conduction velocity is only slightly or not at all affected while the action potential is reduced (32, 56). As a matter of fact, a decrease in amplitude and a change of the shape of the nerve action potential is considered to be the most sensitive sign of neuropathy (48, 68).

In our investigation, the action potential proved to be a less sensitive index of nerve lesion, possibly because we used surface electrodes instead of needle electrodes. Examination with needle electrodes is painful and would have rendered the recruitment of subjects more difficult. Presumably, surface electrodes give a larger scatter of the measured amplitude values, and thereby the sensitivity is lowered. The figures in table 10 lend some support to this assumption. The differences in the amplitudes of the nerve action potentials between the exposed and reference groups were proportionally greater than the differences in conduction velocities, but they were nevertheless not statistically significant due to the great dispersion of the values.

The observed slight decrease in conduction velocity in the exposed group (table 10) is consistent with axonal neuropathy. It has been considered that the primary changes in the axon — with, eg, paranodal swelling — causes a secondary degeneration of the myelin so that the conduction velocity is affected (48). Also in other studies on workers with long-term exposure to organic solvents, an affection of the nerve conduction velocity has been found (3, 67, 75). In our study, the observed differences between the exposed workers and the referents were in general the most pronounced in the lower age groups. This result can depend on the healthy worker effect (57), which means that those whose health deteriorates as a result of influence of the occupational environment leave their work for another type of work. Consequently, a majority of the remaining workers — in higher age groups — have suffered no or only slight damage due either to higher resistance or to other factors.

Vibratory perception thresholds

The exposed group had significantly higher vibration thresholds than reference group 2. A vibration threshold examination tests the function of large sensory nerve fibers. In clinical work, impaired vibration sensibility has long been regarded as an early sign of peripheral neuropathy. Lindblom & Goldberg (50) found, in an examination of jet fuel-exposed workers, that vibration thresholds were more sensitive than ENeG as a test of peripheral neuropathy.

According to Schaumburg & Spencer (72) light touch, pain ("pricking pain"), and temperature discrimination are more reduced than the vibration sense in cases of toxic occupational neuropathies. This possibility was not examined in our investigation, but the findings are unexpected since it is well known that the large fibers are those first affected while pain and temperature sensations are transmitted by small fibers.

An increased vibration threshold is not necessarily associated with sensory symptoms. On the other hand, it cannot be ruled out that an increased threshold can be related to impaired performance in tests requiring good tactile discrimination.

Conclusion

In conclusion, we wish to stress that the exposed group consisted of people in full daily work. A priori we do not have any reason to expect an overrepresentation of definitely pathological findings. Nevertheless, we found statistically significant differences between the exposed group and, in particular, reference group 2 that indicated a reduced function of peripheral nerves in the exposed group. However, the differences were small, and all the mean values were within normal limits. It is known that ENeG changes indicative of peripheral neuropathy can be found in patients (with, eg, diabetes, pernicious anemia, uremia) without symptoms or signs of sensory disturbances. The neurophysio-logical findings from our subjects may therefore not be associated with obvious functional disturbances during ordinary daily activities.

NEURORADIOLOGICAL EXAMINATION

When this investigation was initiated, work in our department had recently indicated that degenerative changes of the brain of alcoholics can easily be detected by computed tomography (CT scanning) (60). This fact aroused an interest to determine whether changes of the same kind can be observed in industrial painters. CT scanning has the great advantage of being noninvasive, and the radiation dose is low, approximately of the same magnitude as that of an ordinary skull radiograph. The method is therefore, in contrast to pneumoencephalography, well suited for investigating large series of individuals.

Material and methods

The method and its theoretical principles will not be dealt with in detail. The reader can consult, for instance, Ambrose (4) and Hounsfield (41) for this information. Basically, an image of a slice of the brain (tomogram) is produced by computer calculations of data obtained from X-ray transmission recordings in the transversal plane. An accurate delineation of the anatomic structures is obtained, and at the same time the machine produces reliable values of the X-ray attenuation of the tissues incorporated in the slice.

An EMI-Mark I head unit was used. The whole brain was examined, usually requiring eight to ten tomographic cuts, each cut being 13-mm thick. The cuts were oriented parallel to the supraorbitomeatal plane. The pictures produced were composed of 160×160 picture elements.

The analysis of the CT scans was carried out on pictures obtained from the same terminal and all with the same window width and level of gray scale. Because of faulty tapes and other technical errors pictures with these specifications were not obtained for 18 individuals (evenly distributed in the different groups), and these CT scans were consequently not used for the analysis of cortical changes or for direct measurements.

Furthermore all analyses were performed after a thorough random mixing of the individual scans and without access to any identification data whatsoever, ie, completely blindly. With the exception of cortical changes, which are reported for three separate investigators, no parameter was measured or estimated by more than one person.

Results

The analysis comprised three different parts. Part 1 consisted of an analysis of the frequency of calcifications and the incidence of local abnormalities in the shape of the ventricles and the cisterns (the fluid-filled cavities within and surrounding the brain). These parameters are listed in table 11. No significant differences could be demonstrated between the exposed and reference groups.

Part 2 comprised measurements carried out with vernier callipers directly from the CT scans of certain anatomic structures. The mean values of these variables, corrected for reduction in the scan, are listed in table 12, which also includes a calculated anterior horn index (51). As seen from the table, a relatively large number of values were missing for the cerebral fissures. Often these structures cannot be delineated accurately with the technique used. However, since the number of missing values is of the same magnitude for the different groups, these mean values have been considered representative and have been included in fig 19, where the calculated differences for all linear variables are given as Z values. As seen from the figure, only one significant ($p < 0.05$) difference was obtained, ie, for inner transverse skull diameter (between the exposed group and reference group 1). It should be noticed, however, that all the mean values for the ventricular parameters were higher for the reference groups though the differences did not reach statistically significant levels.

The third part of the analysis consisted of an assessment of the cortical sulci, made by three independent investigators (two neuroradiologists, investigators 1 and 2 in fig 19, and one laboratory assistant with great experience with CT scan analysis, investigator number 3 in fig 19). The following rating scale was used: 1 = no

Table 11. Incidence of computed-tomography findings. (WTS = wide frontolateral sulci suggesting impeded cerebrospinal fluid circulation, PC = calcification of pineal gland, GCS = calcification of left glomus, GCD = calcification of right glomus, AL = calcification with other location, VFV = visualization of fourth ventricle, VTHS = visualization of left temporal horn, VTHD = visualization of right temporal horn, VCBC = wide cisterns in the posterior fossa)

Group	WTS	PC	GCS	GCD	AL	VFV	VTHS	VTHD	VCBC
Exposed	7	66	47	50	4	44	0	0	9
Reference 1	6	60	37	38	7	45	4	3	10
Reference 2	6	70	52	52	10	51	2	2	8

Table 12. Mean values of linear parameters of the computed-tomography analysis for the different groups.

Group	Left anterior horn Number	Left anterior horn Mean	Right anterior horn Number	Right anterior horn Mean	Anterior horn index Number	Anterior horn index Mean	Largest inner skull diameter Number	Largest inner skull diameter Mean	Transverse diameter 3rd ventricle Number	Transverse diameter 3rd ventricle Mean	Left sylvian fissure Number	Left sylvian fissure Mean	Right sylvian fissure Number	Right sylvian fissure Mean	Interhemispheric fissure Number	Interhemispheric fissure Mean
Exposed	73	16.95	73	16.29	73	0.26	73	130.12	62	4.40	30	4.62	27	5.07	48	6.59
Reference 1	70	17.01	70	16.79	69	0.27	70	128.41	60	4.71	27	4.67	24	4.92	49	6.42
Reference 2	74	17.39	74	17.14	74	0.27	74	129.57	70	4.77	39	4.88	46	5.11	51	5.96

changes = no widened sulci [6]; 2 = suspected degenerative changes = five to ten widened sulci in at least two tomographic cuts; 3 = clear-cut degenerative changes = wide sulci in most cuts; 4 = high grade degenerative changes = markedly widened sulci appearing in all lobes.

The ratings according to the classification were treated as continuous variables, and the results have been reported as mean values for each individual investigator (table 13). Calculated differences between the groups are reported as Z values (fig 19, lower part).

For two of the investigators there was a significant difference (p < 0.05) between the exposed group and reference group 2. The difference was negative (directed to the left in fig 19), which means a higher rating for the reference group. From table 13 it is furthermore obvious that the rating values were generally higher for the two reference groups.

An analysis of the ratings for different age strata in the exposed group and reference group 2 showed that the painters in the youngest age group generally had somewhat higher values than the referents, whereas the reverse was true for the older age groups. Furthermore, there was a tendency towards an increasing deviation between the groups with increasing age. For all three investigators, the difference between the painters and reference group 2 was the greatest in the oldest age group (56—65 a).

Discussion

Recent neuroradiological investigations have given evidence of brain atrophy in solvent-exposed workers with clinical cerebral symptoms (7, 36, 42).

The results obtained in our investigation showed certain differences in a direction opposite to those just referred to, ie, our results seem to indicate that our exposed workers had an increased relative brain volume when compared to the referents. It should be emphasized, however,

that the painters of the present study were clinically healthy and working when the examinations were undertaken, whereas the exposed workers of the other investigations referred to had been given neuroradiological examinations because of severe cerebral symptoms related to solvent exposure. Most of them were either retired on a disability pension or had had to change jobs.

One can only speculate as to the possible mechanisms behind our findings. The reason for an increased brain volume can be brain edema reflecting a low grade toxic effect of the organic solvents. It is, of course, not possible to decide whether the observed changes are reversible or not or whether they may pass into a stage with more degenerative characteristics.

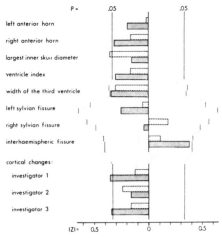

Fig 19. Differences in the mean values of the linear computed-tomography variables of the exposed and reference groups expressed as Z values. Light bars represent reference group 1, and dark bars reference group 2.

Table 13. Rating of the cortical changes. Mean values for the different groups and different investigators.

Group	N	Investigator 1	Investigator 2	Investigator 3
Exposed	73	1.73	1.35	1.37
Reference 1	70	1.86	1.54	1.47
Reference 2	73	2.05	1.47	1.58

---

[6] A sulcus is considered wide when its estimated width exceeds 1 mm in the CT scan, approximately corresponding to a real width of 4 mm.

Finally it should be mentioned that no consistent differences in the width of the Sylvian and intrahemispheric fissures were demonstrated between the groups, and the results do not indicate any derangement of the cerebrospinal fluid circulation in the exposed group.

nish epidemiologic study published by Raitta et al in 1976 (66). They found more lens opacities in a group of car painters than in a reference group of railway engineers, matched with respect to age. Because of this observation an ophthalmologic examination was included in the present study, special attention being paid to the possible occurrence of lens opacities.

OPHTHALMOLOGIC INVESTIGATION

It is well known that high concentrations of vapors from organic solvents can cause an acute transient irritation of the eyes with conjunctival redness and tearing. Some case reports also describe changes of the retina and the optic nerve in workers exposed to organic solvents (35, 86). It is likely that the levels of exposure have been relatively high, certainly higher than the levels measured in contemporary workshops. With these exceptions, no studies suggesting injury to the eyes from occupational exposure to organic solvents have been published until recently. Much attention was therefore drawn to a Fin-

Method

The eye examinations were performed by two ophthalmologists. The first 56 persons were examined and the findings evaluated by both ophthalmologists together. The following 14 persons were also examined by both ophthalmologists, but the findings were evaluated independently. The evaluation of the findings of the two ophthalmologists showed good reliability, and therefore the remaining 168 persons were examined only by one of the two. The workers came in pairs so that one exposed person and one person out of the first reference group were examined on the same occasion by the same doctor. The exposure

Table 14. Contents of the examination of the lens.

Item number	Object of examination	Evaluation
1	Capsule: Presence of coating material on or changes in the anterior lens capsule	Graded: 1 = absence, 2 = presence
2	Anterior subcapsular opacities (except vacuoles)	Graded from 1 to 5
3	Anterior subcapsular vacuoles	Graded: 1 = no vacuoles, 2 = one small vacuole, 3 = more than one small or one big vacuole
4	Posterior subcapsular opacities	Graded from 1 to 5
5	Posterior subcapsular vacuoles	Graded: 1 = no vacuoles, 2 = one small vacuole, 3 = more than one small vacuole or one big vacuole
6	Zones of discontinuity, zones with a slightly increased light scattering in the superficial lens cortex	Counted
7	Punctate opacities in the lens cortex	Counted in a 0.2-mm wide central slit light beam
8	Larger cortical opacities	Graded after their extension
9	Light scattering in the deepest part of the anterior cortex, the supranuclear zone	Graded from 1 to 5
10	Transparency of the nucleus	Graded from 1 to 5
11	Color of the nucleus	Graded from 1 to 5
12	Punctate opacities close to the suture lines in the nucleus	Graded: 1 absence, 2 = presence

was not known to the examiner. The second reference group was examined separately.

The corrected visual acuity of each eye was determined. The pupils were dilated to 7—8 mm with 1 % Cyklogyl® and 10 % Neosynefrine®. The anterior segment of the eye was examined with a Haag-Streit 900 slit lamp microscope. The central fundus was inspected with direct ophthalmoscopy. The intraocular pressure was measured with applanation tonometry in persons over 40 a of age. The slit lamp examination comprised, in particular, a thorough examination of the lens. The lens opacities found were photographed with a Nikon photo slit lamp and marked in a schematic drawing of the lens. The form used for the lens examination is shown in table 14.

Results

The occurrence of clinical pathological ophthalmologic findings was of the same proportion among both the exposed and the nonexposed workers (table 15). The suspected or clearly pathological lens changes are presented in table 16. There was no predominance of lens changes in the exposed group. The lens opacities were in no case big enough to cause reduced visual acuity. The intraocular pressure in persons over 40 a of age was of the same order of magnitude in all the three groups.

The statistical analysis of the lens data and the visual acuity showed only small differences between the exposed and reference groups (fig 20). The number of posterior subcapsular vacuoles was slightly higher in reference group 1 than in the exposed group (see also table 16). The diffuse opacity and the yellow color of the nucleus were more pronounced (p < 0.05) in reference group 2 than in the exposed group. Punctate opacities close to the suture lines in the lens nucleus showed significant differences between the exposed and both of the nonexposed groups. These opacities were found more frequently in the exposed persons than in the nonexposed (p < 0.05 and < 0.001 for reference groups 1 and 2, respectively).

When the material was divided into age groups, no additional significant differ-

Table 15. Number of persons in the examined groups with pathological eye changes.

Pathological finding	Exposed group	Reference group 1	Reference group 2
Amblyopia of one eye	5	1	1
Keratoconus		1	
Central chorioretinal scars		2	
Degenerative maculopathy	1	2	1
Diabetic retinopathy	1		1
Congenital atrophy of the optic nerve	1	1	
Glaucoma	1		
Grave myopia			1

Table 16. Number of persons in the examined groups with suspected or clearly pathological lens changes in at least one eye.

Pathological finding	Exposed group	Reference group 1	Reference group 2
Anterior subcapsular opacities			1
Anterior subcapsular vacuoles	1	1	2
Posterior subcapsular opacities	3	5	2
Posterior subcapsular vacuoles	11	18	16
Small cortical opacities (except punctate)	11	16	20
Larger cortical opacities	5	6	6
Punctate opacities in the nucleus	43	29	20

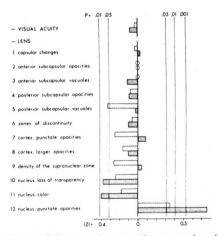

Fig 20. Differences between the exposed and reference groups in regard to the mean values of visual acuity and lens parameters according to the description in table 15. The differences are expressed in Z values. Light bars refer to reference group 1, and dark bars to reference group 2.

ences were found between the exposed and reference groups. However, in the whole material, there was, as could be expected, a distinct correlation between age and some parameters. With increasing age, visual acuity was reduced, the number of opacities in the lens cortex increased, the degree of light scattering in the supranuclear zone increased, the transparency of the nucleus decreased, and its yellow-brown pigmentation increased.

Discussion

Toxic cataract is primarily known as a side effect of treatment with corticosteroids (13) and cholinesterase inhibitors (10, 64). The drugs that were first described to cause cataracts were dinitrophenol and dinitrochresol. These drugs were given for reducing purposes in the 1930s and were found to cause a cataract in 1 % of the persons treated (40). Other substances that have been reported to cause cataracts in humans are triparanol (43), a cholesterol-lowering drug now withdrawn from the market, naphthalene (31), myleran (65), and phenothiazines (78).

Recently, organic solvents used in car paint have been strongly suspected to induce lens changes (66). Nau et al (62) have shown experimentally that the inhalation of high concentrations of some fractions of petroleum distillates can cause cataracts in rats. This result must be interpreted with a certain amount of caution as the general condition of the animals was poor at the onset of cataract.

The first lens change due to toxic influence on the lens is very frequently the appearance of ṣubcapsular vacuoles and subcapsular opacities. Disturbed function of the lens epithelium has been shown experimentally to be an important pathogenic factor in toxic cataracts (69).

In the present study, all visible subcapsular changes and opacities in the cortex and the nucleus of the lens were registered. The diffuse light scattering in the zones of discontinuity, the supranuclear zone, and the nucleus was also estimated. The mean values for the diffuse opacity and yellow color of the nucleus in reference group 2 were 0.14 and 0.16 units higher, respectively, than for the

exposed group. This grading from 1 to 5 was performed without any objective reference system. Consequently it contains a relatively large subjective factor which is minimized when one exposed person and one referent are examined at the same time. This was the case for reference group 1. Reference group 2 was examined later and the larger difference in values for diffuse scattering and yellow color of the nucleus can possibly be explained by a small change in the grading scale.

The most likely effect of organic solvents on the nucleus would have been opacification. This is a known effect of dimethylsulfoxide (DMSO) which, after oral intake, causes an increased nuclear density in several experimental animals (46). The only parameter for which a significant difference was found with respect to both reference groups was microopacities in the suture area of the nucleus (fig 20).

These opacities, due to their position in the lens, can be expected to be congenital changes. However, the distinct predominance for these opacities in the exposed group (table 16) makes it obvious that an effect of the solvents cannot be excluded. The microopacities were vacuoles close to the sutures, ie, at the ends of the lens fibers. Fagerholm (23) has shown experimentally that the ends of lens fibers are sensitive to changes in the ionic environment, resulting in formation of vacuoles. The influence of organic solvents may cause osmotic disturbances in the central parts of the lens and lead to a similar formation of vacuoles.

Our results differ from the results of the Finnish study (66) in which significantly more lens opacities were found among car painters than among referents when age-matched pairs were compared.

In our study a similar comparison of pairs composed of one person from the exposed group and one from reference group 1 produced 18 pairs in which the exposed person had more lens opacities, 23 pairs in which the nonexposed had more opacities, and 35 pairs without any observed difference. When reference group 2 was used for the same comparison, the following results were obtained: in 24 pairs the exposed person had more lens changes, in 24 pairs the nonexposed had more, and in 28 pairs no difference was found. These numbers do not show any significant differences. In our own, as well as in Raitta's Finnish investigation, the small opacities in the center of the lens were not considered. The Finnish reference group consisted of railway engineers, who are a group with high demands on vision, a fact which might influence the result. Other differences between the two investigations might be the method of documenting minimal lens opacities.

To summarize, we have found an increased frequency of microopacities in the central parts of the lenses of the exposed persons. On the other hand, there were no indications that prolonged exposure to solvent vapors, at the levels in question, causes reduced vision or induces cataract formation. Acute effects of solvents on the conjunctiva and the cornea were not studied.

SUMMARY

As indicated in the introductory sections, the purpose of this investigation was to ascertain whether a group of workers with long-term occupational exposure to solvents differed from other, analogous groups of unexposed workers in regard to the nervous system and sensory functions. Both the car painters and industrial painters, as well as the industrial workers in the reference groups, were fully fit for work, and there was no reason to expect any grave clinical abnormalities among them. It was therefore necessary to use methods which, as far as was possible, would permit differentiation within the clinically "normal" range.

The results from the occupational hygiene study yielded a time-weighted mean value of 0.21 for the hygienic effect for both the car and industrial painters. The same mean value for two individuals' need not imply the same biological effect: a steady exposure to levels near the mean value could have another (and perhaps milder) effect than exposure to extremely high levels alternating with low levels, or periods of no exposure. It should also be noted that, when the hygienic effect is being evaluated, a purely additive effect from the various solvents is presumed. However, a potentiating synergism among

different solvents cannot be ruled out. These problems were not examined in our investigation.

The painters were also exposed to lead, although usually in relatively low doses (exposure levels below one-fifth of the accepted limit). The industrial painters were considered to have been less exposed to lead than the car painters; this has definitely been the case at any rate over the past 20 a. Since the same effects were found among younger industrial workers as among the group of painters as a whole, it is our view that solvent exposure has been the most important factor responsible for the observed difference between the exposed group and the reference groups.

It has recently been demonstrated that certain solvents have a relatively long half-time (days) as regards elimination from the human body (22). The painters in our study were definitely unexposed to solvents for only 18—24 h prior to examination. We therefore do not know to what extent the observed effects were or were not reversible. Presumably, several months without exposure would be necessary to shed light on this question. However, the question is but of secondary importance for standard setting (TLV). As long as painters work under exposure to the solvent concentrations found in this investigation they are affected, regardless of whether the symptoms are reversible or not.

We approached the problem of the neurotoxic effect of occupational exposure to solvents in this study from a number of perspectives, employing methods from various disciplines. It should be stressed that the different results all point in the same direction. In the questionnaires and the controlled psychiatric evaluations, the exposed subjects exhibited more neurasthenic symptoms than the reference subjects; they performed more poorly on psychological tests of memory, manual dexterity, perceptual speed, and reaction capability; and in neurophysiological tests they had slower nerve conduction velocities, lower nerve action potentials, and higher vibratory perception thresholds. In addition the results from computed tomography and eye examinations revealed certain differences among the groups.

Since the study was designed so that

256

only exposure to solvents should distinguish the exposed group from the reference groups and since lead exposure was not felt to be an adequate explanation for the observed differences among the groups, the conclusion seems to be warranted that solvent exposure was the cause of these differences.

We have not been able to demonstrate a correlation between degree of exposure and extent of effect (dose-response correlation). The measure of exposure defined proved to have a very high correlation with age. Some correlation between age and effect was demonstrated in some of the studies although not without ambiguity. Indeed, it is probably very difficult in an investigation of this nature to ascertain a dose-response relationship on account of the healthy worker effect.

In conclusion, it should be pointed out that the measured solvent concentrations — both the current levels and the levels from the reconstructed model of a workplace of 25 a ago — were less than half of the TLV of the current Swedish standard. The same was true for Finnish car painter studies (see the introduction).

REFERENCES

1. Åsberg M, Perris C, Schalling D, Sedvall G. A comprehensive psychopathological rating scale. Acta psychiatr scand suppl 271 (1978) 5—28.
2. Åstrand I, Gamberale F. Effects on human beings of solvents in the inspiratory air: A method for estimation of uptake. Environ res 15 (1978) 1—4.
3. Allen N, Mendell JR, Billmaier DJ, Fontaine RE, O'Neill J. Toxic polyneuropathy due to methyl n-butyl ketone. Arch neurol (Chicago) 32 (1975) 209—218.
4. Ambrose J. Computerized transverse actial scanning (tomography): Part 2, clinical application. Br j radiol 46 (1973) 1023.
5. American Conference of Governmental Industrial Hygienists. Threshold limit values for chemical substances and physical agents in the workroom environment. Cincinatti, OH 1978.
6. Andersson NA. Scales and statistics: Parametric and nonparametric. Psychol bull 58 (1961) 305—316.
7. Arlien-Søborg P, Bruhn P, Gyldensted C, Melgaard B. Chronic painter's syndrome: Chronic toxic encephalopathy in house painters. Acta neurol scand 60 (1979) 149—156.
8. Asbury AK. Pathology of industrial toxic neuropathies. Acta neurol scand (1979): suppl 73, 52—53.
9. Axelson O, Hane M, Hogstedt C. A case referent study on neuropsychiatric disorders among workers exposed to solvents. Scand j work environ health 2 (1976) 14—20.
10. Axelsson U. Glaucoma miotic therapy and cataract. Studies on echothiophate (Phospholinen iodine) and Para-Oxon (Mintachol) with regard to cataractogenic effect. Acta ophthalmol suppl 102 (1969).
11. Bederoff-Petersson A, Jägtoft B, Åström J. EPI; Eysenck personality inventory: Comments and results of some Swedish investigations [in Swedish]. Skandinaviska Testförlaget, Stockholm 1968.
12. Bergström I, Lindblom U, Norée L-O. Preservation of peripheral nerve function in severe uremia during treatment with low protein high caloric diet and surplus of essential amino acids. Acta neurol scand 51 (1975) 99—109.
13. Black LR, Oglesby RB, von Sallman L, Bunim JJ. Posterior subcapsular cataracts induced by corticosteroids in patients with rheumatoid arthritis. J am med assoc 174 (1960) 166—171.
14. Browning E. Toxicity and metabolism of industrial solvents. Elsevier Publishing Co, Amsterdam, London, New York 1965.
15. Caruso G, Santoro L, Perretti A, Serlenga L, Rossi L. Electrophysiological findings in the polyneuropathy of leather cementing workers. Acta neurol scand (1979): suppl 73, 61—62.
16. Catell RB, Kline P. The scientific analysis of personality and motivation. 2nd ed. Academic Press New York, London, San Francisco 1977, pp 32—36.
17. Cavanagh JB. Peripheral neuropathy caused by chemical agents. CRC crit rev toxicol 2 (1973) 365—417.
18. Cianchetti C, Abbritti G, Perticoni G, Siracusa A, Curradi A. Toxic polyneuropathy of shoe-industry workers: A study of 122 cases. J neurol neurosurg psychiatry 39 (1976) 1151—1161.
19. Edwards AL. Techniques of attitude scale construction. Appleton-Century-Croft. New York, NY 1967, pp 149—171.
20. Ekberg K, Berglund M, Friberg S, Kadefors R, Persson L-O, Petersén I, Piros S, Örtegren R. Neurophysiology — solvents [in Swedish]. Laboratory of Clinical Neurophysiology and Department of Occupational Medicine, Sahlgrenska Hospital and AB Götaverken, Göteborg 1968. (ASF-project 77/15).
21. Englund A. Mortality and cancer incidence among Swedish painters. Presented at the International symposium on occupational health hazards encountered in surface coating and handling of paints in the construction industry. Bygghälsan, Stockholm, 3—5 October 1979.
22. Engström J. Organic solvents in human adipose tissue. Arbetarskyddsverket, Stockholm 1978. (Arbete och hälsa no 22).

23. Fagerholm P. The influence of calcium on lens fibres. Exp eye res 28 (1979) 211—222.

24. Frey T. The concept of lesion in psychiatry [in Swedish]. Forsk praktik 8 (1976) 55—63.

25. Frömark A, Gamberale F, Sjöborg Å. A psychological test-battery for behavioural toxicology investigations: Application of a previously standardized battery [in Swedish]. University of Stockholm, Stockholm 1979. (Information from PTI nr 104, 1979).

26. Gamberale F. Behavioral effects of exposure to solvent vapors. Arbetarskyddsverket, Stockholm 1975. (Arbete och hälsa no 14).

27. Gamberale F. Behavioral effects of exposure to solvent vapors: Experimental and field studies. In: Horvath M, ed. Adverse effects of environmental chemicals and psychotropic drugs. Vol 2. Elsevier Scientific Publishing Co., Amsterdam 1976, pp 111—133.

28. Gamberale F, Annwall G, Hultengren M. Exposure to xylene and ethylbenzene: III. Effects on central nervous functions. Scand j work environ health 4 (1978) 204—211.

29. Gamberale F, Lisper HO, Anshelm Olson B. The effects of styrene vapors on the reaction time of workers in the plastic boat industries. In: Horvath M, ed. Adverse effects of environmental chemicals and psychotropic drugs. Vol 2. Elsevier Scientific Publishing Co., Amsterdam 1976, pp 135—148.

30. Gardell B. Production engineering and work satisfaction: A social psychological study of industrial work [in Swedish]. PA-rådets förlag, Stockholm 1971. (PA-rådets meddelande nr 63).

31. Gehring PJ. The cataractogenic activity of chemical agents. CRC crit rev toxicol 1 (1971) 93—118.

32. Gilliatt RW. Recent advances in the pathophysiology of nerve conduction. In: Desmedt JE, ed. New development in electromyography and clinical neurophysiology. Vol 2. Karger, Basel 1973, pp 2—18.

33. Götell P, Axelson O, Lindelöf B. Field studies on human styrene exposure. Work environ health 9 (1972) 76—83.

34. Grandjean E, Munchinger R, Turrian V, Haas PA, Kluepfel H-K, Rosenmund H. Investigations into the effects of exposure to trichloroethylene in mechanical engineering. Br j ind med 12 (1955) 131—142.

35. Grant WM. Toxicology of the eye. 2nd ed. Charles C. Thomas Publishing Co, Springfield, IL 1975, pp 937—938.

36. Gregersen P, Mikkelsen S, Klausen H, Døssing M, Nielsen H, Thygesen P. A chronic cerebral painter's syndrome [in Danish]. Ugeskr laeg 140 (1978): 27, 1638—1644.

37. Hänninen H. Psychological picture of manifest and latent carbon disulphide poisoning. Br j ind med 28 (1971) 374—381.

38. Hänninen H, Eskelinen L, Husman K, Nurminen M. Behavioral effects of long-term exposure to a mixture of organic solvents. Scand j work environ health 2 (1976) 240—255.

39. Hane M, Axelson O, Blume J, Hogstedt C, Sundell L, Ydreborg B. Psychological function changes among house painters. Scand j work environ health 3 (1977) 91—99.

40. Horner WD. Dinitrophenol and its relations to formation of cataract. Arch ophthalmol 27 (1942) 1097—1121.

41. Hounsfield GN. Computerized transverse actial scanning (tomography): Part I. Description of system. Br j radiol 46 (1973) 1016.

42. Juntunen J, Eistola P, Hupli V, Hernberg S. Brain atrophy and exposure to organic solvents. In: Proceedings of the IIIrd industrial and environmental neurology congress. Prague 1979, p 46.

43. Kirby TJ. Cataracts produced by triparanol. Trans am ophthalmol soc 65 (1967) 493—543.

44. Knave B, Anshelm Olson B, Elofsson S, Gamberale F, Isaksson A, Mindus P, Persson HE, Struwe G, Wennberg A, Westerholm P. Long-term exposure to jet fuel: II. A cross-sectional epidemiologic investigation on occupationally exposed industrial workers with special reference to the nervous system. Scand j work environ health 4 (1978) 19—45.

45. Knave B, Persson HE, Goldberg JM, Westerholm P. Long-term exposure to jet fuel: An investigation on occupationally exposed workers with special reference to the nervous system. Scand j work environ health 3 (1976) 152—164.

46. Kuck JF Jr. Drugs influencing the lens. In: Dirkstein S, ed. Drugs and ocular tissues. Karger, Basel 1977.

47. Le Quesne PM. Neurophysiological investigation of subclinical and minimal toxic neuropathies. Muscle nerve 1 (1978) 392—395.

48. Le Quesne PM. Electrophysiological investigation of toxic neuropathies in man. Acta neurol scand (1979): suppl 73, 53—55.

49. Levin M. Occupational hygiene survey of the work environment of car and industry spray painters [in Swedish]. Arbetarskyddsstyrelsen, Stockholm 1977—1978. (Arbetarskyddsstyrelsens undersökningsrapporter 1977: 30, 1978: 16).

50. Lindblom U, Goldberg JM. Screening for neurological symptoms and signs after exposure to jet-fuel. Acta neurol scand (1979): suppl 73, 64.

51. Lindgren E. Encephalography in cerebral atrophy. Acta radiol 35 (1951) 277.

52. Lindström K. Psychological performances of workers exposed to various solvents. Work environ health 10 (1973) 151—155.

53. Lindström K, Härkönen H, Hernberg S. Disturbances in psychological functions of workers occupationally exposed to styrene.

Scand j work environ health 2 (1976) 129—139.

54. Lishman WA. Organic psychiatry: The psychological consequences of cerebral disorder. Blackwell Scientific Publishing Co, London 1978.

55. Marke S, Nyman E. Manual of the MNT-scale [in Swedish]. Skandinaviska Testförlaget AB, Stockholm 1958.

56. McLeod JG. Nerve conduction measurements for clinical use. In: v Duijn H, Donker D, v Huffelen AC, ed. Current concepts in clinical neurophysiology didactic lectures of the ninth international congress of EEG and clinical neurophysiology. Electroenceph clin neurophysiol 43 (1977) 449—636.

57. McMichael AJ, Haynes SG, Tyroler HA. Observations on the evaluation of occupational mortality data. J occup med 17 (1975) 128—131.

58. Mikkelsen S. A cohort study of disability pension and death among painters with special regard to disabling presenile dementia as an occupational disease. Scand j soc med suppl 16 (1980) 34—43.

59. Mikkelsen S, Gregersen P, Klausen H, Døssing M, Nielsen H. Presenile dementia as an occupational disease following industrial exposure to organic solvents: A review of the literature [in Danish]. Ugeskr laeg 140/27 (1978) 1633—1638.

60. Myrhed M, Bergman H, Borg S, Hindmarsh T, Ideström C-M. Computer tomographic cerebral findings in a group of alcoholic patients. Hygiea acta soc mediocrum sueciana 85 (1976) 253.

61. National Board of Occupational Safety and Health. Direction no 100. Solna 1978.

62. Nau CA, Neal J, Thornton M. C_9—C_{12} fractions obtained from petroleum distillates. Arch environ health 12 (1966) 382—393.

63. Nilsson B. Visual evoked responses in multiple sclerosis comparison of two methods for pattern reversal. J neurol neurosurg psychiatry 41 (1978) 499—504.

64. Philipson B, Kaufman P, Fagerholm P, Axelsson U, Bárány EH. Electron microscopy and microradiography of echothiophateinduced cataracts in cynomolgus monkeys. Arch ophthalmol 97 (1979) 340—346.

65. Podos SM, Canellos GP. Lens changes in chronic granulocytic leukemia: Possible relationship to chemotherapy. Am j ophthalmol 68 (1969) 500—504.

66. Raitta CH, Husman K, Tossavainen A. Lens changes in car painters exposed to a mixture of organic solvents. Albrecht von Graefes Arch Klin Exp Ophthalmol 200 (1976) 149—156.

67. Rosén I, Aronsen B, Rehnström S, Welinder H. Neurophysiological observations after chronic styrene exposure. Scand j work environ health 4 (1978): suppl 2, 184—194.

68. Rosenfalck A. Early recognition of nerve disorders by near-nerve recording of sensory action potentials. Muscle nerve 1 (1978) 360—367.

69. von Sallman L. The lens epithelium in the pathogenesis of cataract. Am j ophthalmol 44 (1957) 159—170.

70. Schaumburg HH. Neurology of toxic neuropathies. Acta neurol scand (1979): suppl 73, 50—51.

71. Schaumburg HH, Spencer PS. The neurology and neuropathology of the occupational neuropathies. J occup med 18 (1976) 739—742.

72. Schmidt R, Schnoy M, Wagner HM, Altenkirch H. Ultrastructural studies on toxic effects in lung tissue induced by inhalation of solvents. Presented at the Second international congress on toxicology, Brussels 6—11 July 1980.

73. Seppäläinen AM. Applications of neurophysiological methods in occupational medicine: A review. Scand j work environ health 1 (1975) 1—14.

74. Seppäläinen AM, Härkönen H. Neurophysiological findings among workers occupationally exposed to styrene. Scand j work environ health 3 (1976) 140—146.

75. Seppäläinen AM, Husman K, Mårtenson C. Neurophysiological effects of long-term exposure to a mixture of organic solvents. Scand j work environ health 4 (1978) 304—314.

76. Seppäläinen AM, Raitta C, Huuskonen S. n-hexan-induced changes in visual evoked potentials and electroretinograms of industrial workers. Electroencephalogr clin neurophysiol 47 (1979) 492—498.

77. Seppäläinen AM, Tolonen M. Neurotoxicity of long-term exposure to carbon disulphide in the viscose rayon industry: A neurophysiological study. Work environ health 11 (1974) 145—153.

78. Sidall JK. The ocular toxic findings with prolonged and high dosage chlorpromazine intake. Arch ophthalmol 74 (1965) 460—464.

79. Siegel S. Non-parametric statistics for the behavioural sciences. McGraw Hill, New York, NY 1956.

80. Sjöborg Å, Frömark A. Standardization of a psychological test battery for behavioral toxicology investigations [in Swedish]. University of Stockholm, Stockholm 1977. (Information from PTI no 93, 1977).

81. Steby M. Occupational hygiene survey of the work environment of car and industry spray painters [in Swedish]. Arbetarskyddsstyrelsen, Stockholm 1977, 1978. (Arbetarskyddsstyrelsens undersökningsrapporter 1977: 10, 18, 21, 28, 29, 33, 1978: 2, 10).

82. Steby M, Levin M. Exposure to organic solvents, dust and metals of car spray painters: A study with reference to earlier work environment conditions [in Swedish]. Arbetarskyddsverket, Stockholm 1979. (Arbete och hälsa no 3).

83. Struwe G. The neurasthenic syndrome [in Swedish]. Läkartidningen 47 (1979) 4253—4256.

84. Wennberg A, Zetterberg L. Application of a computer-based model for EEG analysis. Electroencephalogr clin neurophysiol 31 (1971) 457—468.
85. Westlander G. Work conditions and leisure time contents: A social psychological study of male individual workers [in Swedish]. PA-rådets förlag, Stockholm 1976. (PA-rådets meddelande no 67).
86. Zagora E. Eye injuries. Charles C Thomas Publisher, Springfield, IL 1970, pp 345—371.

Received for publication: 29 September 1980

ADVANCES IN MODERN ENVIRONMENTAL TOXICOLOGY

VOLUME III
Assessment of Reproductive and Teratogenic Hazards

Edited by M.S. CHRISTIAN, Argus Research Laboratories, Inc.
W.M. GALBRAITH, U.S. Environmental Protection Agency
P. VOYTEK, U.S. Environmental Protection Agency
and M.A. MEHLMAN, Mobil Oil Corporation

SECTION I

The 1980's: An Era of Reproductive Confrontation. *J.E. Gocke*
Statement of Problem: Reproductive Hazard. *M.S. Christian*
The Teratologist as a Consultant. *E.M. Johnson*
Pharmaceuticals, Drugs and Birth Defects. *R.M. Hoar*
Food, Food Additives and Natural Products. *G. Nolan*
Practical Applications of Systems for Rapid Detection of Potential Teratogenic Hazards. *E.M. Johnson*
Reproductive Toxicology: Radiation Effects. *R.P. Jensh*
Assessment of Reproductivity Toxicity—State of the Art. *M.S. Christian*
Petroleum and Petroleum Products: A Brief Review of Studies to Evaluate Reproductive Effects. *C.A. Schreiner*

SECTION II

Assessment of Risks to Human Reproduction and to Development of the Human Conceptus from Exposure to Environmental Substances. Proceedings of U.S. Environmental Protection Agency—1982. Documented information by over 100 noted experts in the field of developmental biology and teratology.

Chapter 1: Introduction
Chapter 2:
Female Reproduction
General Reproductive Toxicity Screen
Qualitative Reproductive Toxicity Tests
Quantitative Reproductive Toxicity Tests
Risk Assessment
Research Needed
  Extrapolation of animal data to humans
Details of Test Protocols and Glossary of Terms for Female Risk Assessments
Description and Discussion of Tests Useful in Assessing Risk to the Female Reproductive System
Qualitative Reproductive Toxicity Tests
Quantitative Reproductive Toxicity Tests
Ovarian Toxicity
Chapter 3
Considerations in Evaluating Risk to Male Reproduction

Introduction
Aspects of the Problem
Selection of an Animal Model
Tests for Evaluating Reproductive Damage
Evaluation of Reproductive Damage in Exposed or Potentially Exposed Men
Protocols for Testing Compounds with Animal Models Research Needed
Details of Test Protocols and Glossary of Terms for Male Risk Assessment
Description and Discussion of Tests Useful in Animal Models or Man
Testicular Characteristics
Epididymal Characteristics
Assessment of Male Reproductive Toxicity Using Endocrinological Methods
Examination of Known Toxic Exposures
  Humans
  Animal Models
Fertility Testing
  Tests available
Sperm Nucleus Integrity
  Spematozoal morphology
  Karyotyping of human spermatozoa by the denuded-hamster-egg technique
Dose Response
Chapter 4
Current Status of, and Consideration for, Estimation of Risk to the Human Concepts from Environmental Chemicals
Definition and Scope
Impact of Developmental Abnormalities on Humans
Causes of Congenital Malformations
Qualitative Evaluation of Risk Potential
Chapter 5
Other Considerations: Epidemiology, Pharmacokinetics, and Sexual Behavior
Epidemiology: Methods and Limitations
  Possible data sources and useful approaches
Pharmacokinetics
Sexual Behavior
  Qualitative evaluation of risk potential
  Animal studies
  Assessment of human sexual behavior: surveillance and epidemiological studies

Princeton Scientific Publishing Co., Inc.

ADVANCES IN MODERN ENVIRONMENTAL TOXICOLOGY

VOLUME VI
Applied Toxicology
of Petroleum Hydrocarbons

Edited by H.N. MACFARLAND, Gulf Oil Corporation
C.E. HOLDSWORTH, American Petroleum Institute
J.A. MACGREGOR, Standard Oil of California
and M.L. KANE, American Petroleum Institute

Comparison of the Carcinogenic Potential of Crude Oil and Shale Oil. *R.M. Coomes and K.A. Hazer*

Skin Carcinogenic Potential of Petroleum Hydrocarbons. I. Separation and Characterization of Fraction for Bioassays. *R.W. King, S.C. Lewis, S.T. Cragg, and D.W. Hillman*

Skin Carcinogenic Potential of Petroleum Hydrocarbons. II. Carcinogenesis of Crude Oil, Distillate and Chemical Class Subfractions. *S.C. Lewis, R.W. King, S.T. Cragg, and D.W. Hillman*

Statistical Evaluations in the Carcinogenesis Bioassay of Petroleum Hydrocarbons. *J.M. Holland and E.L. Frome*

API's Approach to Quality Assurance. *R.M. Siconolfi and B.K. Hoover*

Human Sensory Response to Selected Petroleum Hydrocarbons. *L. Hastings, G.P. Cooper, and W. Burg*

Experimental Evaluation of Selected Petrochemicals for Subchronic Neurotoxic Properties. *P.S. Spencer*

Effects of N-Octane Exposure on Schedule-Controlled Responding in Mice. *J.R. Glowa and M.E. Natale*

Inhalation Toxicity of N-Hexane and Methyl Ethyl Ketone. *F.L. Cavender, H.W. Casey, E.J. Gralla, and J.A. Swenberg*

Inhalation Teratology Study of Benzene in Rats. *W.B. Coate, A.M. Hoberman, and R.S. Durloo*

Mechanistic Evaluation of the Pulmonary Toxicology of Nickel Subsulfide. *G.L. Fisher, C.E. Chrisp, K.L. McNeill, D.A. McNeill, C. Democko, and G.L. Finch*

Comparison of the Subchronic Inhalation Toxicity of Petroleum and Oil Shale JP-5 Jet Fuels. *C.L. Gaworski, J.D. MacEwen, E.H. Vernot, R.H. Bruner, and M.J. Cowan*

The Acute Toxicology of Selected Petroleum Hydrocarbons. *L.S. Beck, D.I. Hepler, and K.L. Hansen*

Mutagenicity of Automotive Particulate Exhaust: Influence of Fuel Extenders, Additives, and Aromatic Content. *C.R. Clark, J.S. Dutcher, T.R. Henderson, R.O. McClellan, W.F. Marshall, T.M. Naman, and D.E. Seizinger*

Mutagenicity Evaluation of Petroleum Hydrocarbons. *C.C. Conaway, C.A. Schreiner, and S.T. Cragg*

Percutaneous Absorption of Benzene. *T.J. Franz*

Induction of Xenobiotic Metabolism in Rats on Exposure of Hydrocarbon Based Oils. *A.D. Rahimtula, P.J. O'Brient, and J.F. Paine*

Inhalation Cytogenetics in Mice and Rats Exposed to Benzene. *T.A. Cortina, E.W. Sica, N.E. McCarroll, W. Coate, A. Thakur, and M.G. Farrow*

Consequences Associated with the Inhalation of Uncombusted Diesel Vapor. *R.J. Kainz*

The Toxicity of Petroleum and Shale JP5. *V. Bogo, R.W. Young, T.A. Hill, C.L. Feser, J. Nold, G.A. Parker, and R.M. Cartledge*

Princeton Scientific Publishing Co., Inc.

VOLUME VIII

OCCUPATIONAL AND INDUSTRIAL HYGIENE: CONCEPTS AND METHODS

Edited by Nurtan A. Esmen and Myron A. Mehlman

Chapter I
The Emergence and Evolution of Industrial Hygiene
Anna M. Baetjer

Chapter II
Engineering Control of Occupational Health Hazards
Melvin W. First

Chapter III
Industry-Wide Occupational Health Studies
Robert L. Harris

Chapter IV
On Estimation of Occupational Health Risks
Nurtan A. Esmen

Chapter V
Recent Advances in Respiratory Tract Particle Deposition
Morton Lippmann

Chapter VI
Estimation of Heat Stress Studies
Eliezer Kamon

Chapter VII
Aerosol Measurement
Klaus Willeke

Chapter VIII
Establishing Threshold Limit Values for Airborne Sensory Irritants from an Animal Model and the Mechanisms of Action of Sensory Irritants
Yves Alarie

Chapter IX
The Present State of Occupational Medicine
Bertram D. Dinman

Chapter X
Reacting to Cancer Clusters in the Workplace
Philip E. Enterline

Chapter XI
Pulmonary Clearance
Paul E. Morrow

Appendix
Research Trends and Recent Advances in Occupational Health—
A Selection of Four Papers:
   "Conditions of Work and Man's Health: Tomorrow's Problems" (1965);
   "Significant Dimensions of the Dose-Response Relationship" (1968);
   "Criteria for Hazardous Exposure Limits" (1973);
   "Priorities in Preventive Medicine" (1974).
Theodore F. Hatch